FLUID FLOW PROBLEMS

Edited by **Farhad Ali**

Fluid Flow Problems

http://dx.doi.org/10.5772/intechopen.75996

Edited by Farhad Ali

Assistant to the Editor(s): Nadeem Ahmad Sheikh

Contributors

Taza Gul, Alexandre M. Afonso, Luís Ferrás, Maria L Morgado, Magda Rebelo, Rosalía T. Leiva, António Castelo, Gareth H. McKinley, Krzysztof Mizerski, Han-Yong Jeon, Farhad Ali, Nadeem Ahmad Sheikh

Notice

Statements and opinions expressed in the chapters are these of the individual contributors and not necessarily those of the editors or publisher. No responsibility is accepted for the accuracy of information contained in the published chapters. The publisher assumes no responsibility for any damage or injury to persons or property arising out of the use of any materials, instructions, methods or ideas contained in the book.

First published in London, United Kingdom, 2019 by IntechOpen

IntechOpen is the global imprint of INTECHOPEN LIMITED, registered in England and Wales, registration number: 11086078, The Shard, 25th floor, 32 London Bridge Street
London, SE19SG – United Kingdom
Printed in Croatia

British Library Cataloguing-in-Publication Data
A catalogue record for this book is available from the British Library

Additional hard and PDF copies can be obtained from orders@intechopen.com

Fluid Flow Problems , Edited by Farhad Ali
p. cm.
Print ISBN 978-1-78984-878-6
Online ISBN 978-1-78984-879-3
eBook (PDF) ISBN 978-1-83880-722-1

We are IntechOpen,
the world's leading publisher of
Open Access books
Built by scientists, for scientists

4,200+
Open access books available

116,000+
International authors and editors

125M+
Downloads

Our authors are among the

151
Countries delivered to

Top 1%
most cited scientists

12.2%
Contributors from top 500 universities

CLARIVATE ANALYTICS
BOOK
CITATION
INDEX
INDEXED

WEB OF SCIENCE™

Selection of our books indexed in the Book Citation Index
in Web of Science™ Core Collection (BKCI)

Interested in publishing with us?
Contact book.department@intechopen.com

Numbers displayed above are based on latest data collected.
For more information visit www.intechopen.com

Meet the editor

Dr. Farhad Ali obtained is PhD in Applied Mathematics from one of the world's leading universities, Universiti Teknologi Malaysia. He has more than 13 years of academic experience in different reputed institutions in the country. Currently, he is working as Associate Professor and Head of the Mathematics Department. Additionally, he holds the position of Director ORIC. His experience has mainly been working in the academic sector and he has expertise in fluid dynamics, magnetohydrodynamic flows, nanofluids, fractional derivatives, integral transforms, exact solutions, and mathematical modeling. He has published more than 80 research papers in well-reputed international journals. Apart from academia he has organized academic knowledge seminars and international conferences for students, faculty members, researchers, and practitioners from different parts of the world. Dr. Ali is Head of the Computational Analysis Research Group formed by Ton Duc Thang University, Vietnam. He is the potential reviewer for more than 20 international research journals. He is also a chief editor of *City University International Journal of Computational Analysis* and associate editor for the journal *Advances in Mechanical Engineering*. He is an HEC approved supervisor and has supervised dozens of MS scholars.

Contents

Preface

"A teacher can never truly teach unless he is still learning himself. A lamp can never light another lamp unless it continues to burn its own flame. The teacher who has come to the end of his subject, who has no living traffic with his knowledge but merely repeats his lessons to his students, can only load their minds; he cannot quicken them."

—Rabindranath Tagore

In physics and engineering, fluid dynamics is a subdiscipline of fluid mechanics that describes the flow of fluids, liquids, and gases. It has several subdisciplines, including aerodynamics (the study of air and other gases in motion) and hydrodynamics (the study of liquids in motion). Fluid dynamics has a wide range of applications, including calculating forces and moments on aircraft, determining the mass flow rate of petroleum through pipelines, predicting weather patterns, understanding nebulae in interstellar space and modeling fission weapon detonation. In this book, we provide readers with the fundamentals of fluid flow problems. Specifically, Newtonian, non-Newtonian and nanofluids are discussed. Several methods exist to investigate such flow problems. This book introduces the applications of new, exact, numerical and semianalytical methods for such problems. The book also discusses different models for the simulation of fluid flow.

Chapter 1 is an introductory chapter, providing a brief discussion of fluid flow problems and their application in society.

Chapter 2 is a brief description of existing viscoelastic models, starting with the classical differential and integral models, and then focusing on new models that take advantage of the enhanced properties of the Mittag–Leffler function (a generalization of the exponential function). The generalized models considered in this work are the fractional Kaye–Bernstein, Kearsley, Zapas integral model and the differential generalized exponential Phan-Thien and Tanner (PTT) model. The integral model makes use of the relaxation function obtained from a step-strain applied to the fractional Maxwell model, and the differential model generalizes the familiar exponential PTT constitutive equation by substituting the exponential function of the trace of the stress tensor by the Mittag–Leffler function.

In **Chapter 3**, a magnetohydrodynamic flow of a viscous and conducting fluid confined between two parallel differentially moving boundaries is considered. The whole system is in a strong magnetic field chosen in such a way that the Hartmann boundary layers that form in this problem become singular at the points where the magnetic field becomes tangent to the boundary. Two geometries are taken into account: plane and spherical. Within the class of such configurations the velocity field of the fluid and the influence of the conductivity of the boundaries on the fluid's motion are reviewed here.

The aim of the study in **Chapter 4** is to coat a stretching cylinder with the help of a liquid film spray. Casson fluid has been chosen for the coating phenomena. The thickness of the liquid film has been used variably and the influence of heat and mass transmission under the impact of thermophoresis has been encountered in the flow field. The required pressure term for the spray pattern during variable thickness is the main focus. Using suitable similarity transformations the basic flow equations for fluid motion have been converted into high-order non-linear coupled differential equations. A series of solutions to subsequent problem have been obtained using a controlling procedure optimal approach.

The relationship of compressive behavior according to manufacturing process parameters of GeoNet is investigated in **Chapter 5**. The drainage behavior of the bi-and triplane GeoNet used for planar drainage analyzed and investigated the changes of the drainage behavior due to the restraining load. The data showed that there is no critical manufacturing factor that affects the compressive strength of the biplanar GeoNet. All of these parameters are affected in a very complicated way. The strand inclination mainly affects the after-compressive strength, i.e. roll-over behavior. The results considering site-specific conditions of the landfill system explain that temperature has an influence on the compressive behavior of the GeoNet. Compressive strength was reduced and the strain at yield increased gradually with temperature for both bi- and triplanar GeoNets.

This text is suitable for senior undergraduate students, postgraduate students, engineers, and scientists.

I am grateful to many friends, colleagues, and students around the world who offered their suggestions and help at various stages of the preparation of the book. I express my sincere thanks to my student Mr. Nadeem Ahmad Sheikh, lecturer at City University of Science and IT, Peshawar, for making this project successful. In spite of the best efforts of everyone involved, some typographical errors doubtless remain. Finally, I wish to express my special thanks to the staff of IntechOpen for their help and cooperation.

Dr. Farhad Ali
Head of Department of Mathematics
City University of Science and Information Technology
Peshawar, Pakistan

Introductory Chapter: Fluid Flow Problems

Farhad Ali and Nadeem Ahmad Sheikh

Additional information is available at the end of the chapter

http://dx.doi.org/10.5772/intechopen.81300

1. Introduction

This chapter will cover various flow regimes and their solutions, including, Newtonian, non-Newtonian, and nanofluids via integral transforms and numerical schemes.

1.1. Background

In many real life problems, heat transfer is an important issue and becomes a challenge for the engineers and industrialists. In order to overcome this challenge, one of the methods, which is commonly in use, is to increase the available surface area of heat exchange [1–4]. Hussanan et al. [5] studied the use of oxide nanoparticles for the energy enhancement in water, kerosene, and engine oil-based nanofluids. Tesfai [6] experimentally investigated graphene and graphene oxide suspension for thermal management application. Shafie et al. [7] are considered the first who reported a theoretical study on molybdenum disulfide (MoS_2) nanoparticle suspended in water-based nanofluid in a channel. Khan et al. [8] and Khan [9] also analyzed Molybdenum Disulfide nanofluids in a vertical channel with various effects. Few other interesting investigations in this direction are those made by Wu and Zhao [10], Khan [11], Ali et al. [12] Sheikholeslami and Bhatti [13], Rashidi et al. [14], Mahian et al. [15], and Kasaeian [16].

About 300 years ago, the idea of fractional derivatives was presented [17–21]. This was considered an abstract area of mathematics by many researchers at the initial stages, which will be of no use and will contain only mathematical manipulations. For the last few decades, a new era started in the field of mathematics that changed the interest of scientists from pure mathematics to various applied fields of mathematical sciences, for instance, bioengineering, viscoelasticity, mechatronics and biophysics. Applications of fractional calculus have also been found to be used widely in various fields of science despite mathematics and physics. In fluid dynamics, the noninteger order calculus has been widely used to describe the viscoelastic behavior of the materials. The viscoelasticity of a material is defined as being deformed

IntechOpen

and exhibiting a viscous and elastic behavior through the mechanical energy of storage and simultaneous behavior. The commonly used fractional derivative operators are that of the Riemann-Liouville and the Caputo fractional derivatives. However, there have been some shortcomings in use of these operators. When the Riemann-Liouville fractional derivatives are used, the derivative of a constant is not zero, and some terms are contained without physical significance while applying the Laplace transform, whereas in the case of Caputo fractional derivatives, the kernel is a singular function. To overcome this problem, in 2015, Caputo and Fabrizio have developed a new approach without singularities [18]. The time fractional derivative operator of Caputo-Fabrizio is suitable for the use of the Laplace transform. Frequently, the classical models of equations governing the fluid flow are changed to fractional models, just by replacing derivatives w.r.t time with fractional order derivatives of order $\alpha \in (0, 1)$ see for example [18]. Many scientists and researchers have used the Caputo-Fabrizio fractional derivatives in their studies for physical models [22–28]. Atangana et al. [29] have studied the ground water flowing in aquifer using the applications of the Caputo-Fabrizio derivatives. Very recently, Atangana and Baleanu have presented a new fractional derivative with nonlocal and nonsingular kernel [30–34]. Keeping in mind the above important features, the fractional model for non-Newtonian fluid is considered in the present project.

MHD is the study of magnetic properties of electrically conducting fluids. Liquid metals, plasma, salt water, and electrolytes are the examples of MHD fluid. The pioneering work on MHD has been done by Alfven [35]. In 1970, for his great work, he also received a Nobel Prize. In engineering and technology, MHD has many applications such as hydromagnetic generators (it includes disk system) and MHD flow meters, plasma studies, bearings, pumps, solar energy collection, geothermal energy extractions and nuclear reactors, boundary layer control, extraction of petroleum products, and cooling of the metallic plate. There are many applications of hydromagnetic flow of non-Newtonian fluids in a rotating body in metrology, geographic, turbo machinery, astrophysical, and several other areas. In addition, it has a lot of applications in the biomedical field for instance blood flow in capillaries and flows in blood oxygenation, etc. Also, it has many applications in engineering such as in transpiration cooling, porous pipe design, and design of filters [36]. The role of Hall effect on MHD flow in a rotating frame is remarkable.

In many industrial and natural conditions, the flow through porous media occurs. As rainwater penetrates through the permeable aquifer, hydrological engineering forced flow of oil into sandstone deposits, membrane separation process, drying process and powder technology. Recently, there have been numerous reports dealing with the transport phenomena in porous media, especially due to their importance in various applications, involving the manufacturing and processing industries. It is assumed that the fluid is incompressible, and the fluid flow in the saturated porous medium is treated in most studies where the mass density is constant and the velocity of the fluid is independent of the mass density. The researchers can get help from a better knowledge of free convection through a porous medium in several fields such as heat exchanger, geothermal systems, insulation design, grain storage, catalytic reactors, filtering devices, and metal processing. Recently, attention has been focused on the uses of porous media in high-temperature applications. Porous media are used for the improvement of heat transfer in thermal insulation systems and coolant passages. It is the immeasurable need to ponder on convection flows of Newtonian and non-Newtonian fluids over a vertical oscillating plate passing through a porous medium. In the applied science and engineering, porous media play an important role such as:

- Soil Science: the porous media (soil) contains and transports nutrients and water to plants.

- Hydrology: the porous media are a water bearing and sealing layer.

- Chemical Engineering: porous media are used as a filter or catalyst bed.

- Petroleum Engineering: porous media in the form of reservoir rock, stores, crude oil, and natural gas.

Author details

Farhad Ali[1,2]* and Nadeem Ahmad Sheikh[3]

*Address all correspondence to: farhadali@cusit.edu.pk

1 Computational Analysis Research Group, Ton Duc Thang University, Ho Chi Minh City, Vietnam

2 Faculty of Mathematics and Statistics, Ton Duc Thang University, Ho Chi Minh City, Vietnam

3 Department of Mathematics, City University of Science and Information Technology, Peshawar, Khyber Pakhtunkhwa, Pakistan

References

[1] Yu W, Xie H, Bao D. Enhanced thermal conductivities of nanofluids containing graphene oxide nanosheets. Nanotechnology. 2009;**21**(5):055705

[2] Reddy JR, Sugunamma V, Sandeep N. Impact of nonlinear radiation on 3D magnetohydrodynamic flow of methanol and kerosene based ferrofluids with temperature dependent viscosity. Journal of Molecular Liquids. 2017;**236**:93-100

[3] Choi SUS. Enhancing Thermal Conductivity of Fluids with Nanoparticles. The American Society of Mechanical Engineers: ASME-Publications-Fed; 1995;**231**:99-106

[4] Öztop HF, Estellé P, Yan WM, Al-Salem K, Orfi J, Mahian O. A brief review of natural convection in enclosures under localized heating with and without nanofluids. International Communications in Heat and Mass Transfer. 2015;**60**:37-44

[5] Hussanan A, Salleh MZ, Khan I, Shafie S. Convection heat transfer in micropolar nanofluids with oxide nanoparticles in water, kerosene and engine oil. Journal of Molecular Liquids. 2017;**229**:482-488

[6] Tesfai W, Singh P, Shatilla Y, Iqbal MZ, Abdala AA. Rheology and microstructure of dilute graphene oxide suspension. Journal of Nanoparticle Research. 2013;**15**(10):1989

[7] Shafie S, Gul A, Khan I. Molybdenum disulfide nanoparticles suspended in water-based nanofluids with mixed convection and flow inside a channel filled with saturated porous medium. In: Rusli N, Zaimi WMKAW, Khazali KAM, Masnan MJ, Daud WSW, Abdullah N, et al., editors. AIP Conference Proceedings. Vol. 1775, No. 1. American Institute of Physics: AIP Publishing; 2016. p. 030042

[8] Khan I, Gul A, Shafie S. Effects of magnetic field on molybdenum disulfide nanofluids in mixed convection flow inside a channel filled with a saturated porous medium. Journal of Porous Media. 2017;**20**(5):435-448. DOI: 10.1615/JPorMedia.v20.i5.50

[9] Khan I. Shape effects of nanopartilces on mhd slip flow of molybdenum disulphide nanofluid in a porous medium. Journal of Molecular Liquids. 2017;**233**:442-451. DOI: 10.1016/j.molliq.2017.03.009

[10] Wu JM, Zhao J. A review of nanofluid heat transfer and critical heat flux enhancement-research gap to engineering application. Progress in Nuclear Energy. 2013;**66**:13-24

[11] Khan I. Shape effects of MoS$_2$ nanoparticles on MHD slip flow of molybdenum disulphide nanofluid in a porous medium. Journal of Molecular Liquids. 2017;**233**:442-451

[12] Ali F, Gohar M, Khan I. MHD flow of water-based Brinkman type nanofluid over a vertical plate embedded in a porous medium with variable surface velocity, temperature and concentration. Journal of Molecular Liquids. 2016;**223**:412-419

[13] Sheikholeslami M, Bhatti MM. Active method for nanofluid heat transfer enhancement by means of EHD. International Journal of Heat and Mass Transfer. 2017;**109**:115-122

[14] Rashidi MM, Yang Z, Awais M, Nawaz M, Hayat T. Generalized magnetic field effects in burgers' nanofluid model. PLoS One. 2017;**12**(1):e0168923

[15] Mahian O, Kianifar A, Heris SZ, Wen D, Sahin AZ, Wongwises S. Nanofluids effects on the evaporation rate in a solar still equipped with a heat exchanger. Nano Energy. 2017;**36**:134-155

[16] Kasaeian A, Azarian RD, Mahian O, Kolsi L, Chamkha AJ, Wongwises S, et al. Nanofluid flow and heat transfer in porous media: A review of the latest developments. International Journal of Heat and Mass Transfer. 2017;**107**:778-791

[17] Leibniz GW. Letter from Hanover, Germany, September 30, 1695 to GA l'hospital. JLeibnizen Mathematische Schriften. 1849;**2**:301-302

[18] Caputo M, Fabrizio M. A new definition of fractional derivative without singular kernel. Progress in Fractional Differentiation and Applications. 2015;**1**(2):1-13

[19] Oldham K, Spanier J. The Fractional Calculus Theory and Applications of Differentiation and Integration to Arbitrary Order. Vol. 111. United States of America: Elsevier; 1974

[20] Samko SG, Kilbas AA, Marichev OI. Fractional Integrals and Derivatives. Theory and Applications. Yverdon: Gordon and Breach; 1993

[21] Das S. Functional Fractional Calculus. Berlin, Germany: Springer Science & Business Media; 2011

[22] Magin RL. Fractional Calculus in Bioengineering. Redding: Begell House; 2006

[23] Rossikhin YA, Shitikova MV. Applications of fractional calculus to dynamic problems of linear and nonlinear hereditary mechanics of solids. Applied Mechanics Reviews. 1997;**50**(1):15-67

[24] Carpinteri A, Mainardi F, editors. Fractals and Fractional Calculus in Continuum Mechanics. Vol. 378. London: Springer; 2014

[25] Machado JT, Kiryakova V, Mainardi F. Recent history of fractional calculus. Communications in Nonlinear Science and Numerical Simulation. 2011;**16**(3):1140-1153

[26] Mandelbrot BB. The Fractal Geometry of Nature. San Francisco, CA: Freeman & Co; 1982

[27] Petras I. Fractional-Order Nonlinear Systems: Modeling, Analysis and Simulation. Berlin, Germany: Springer Science & Business Media; 2011

[28] Bagley RL, Torvik PJ. A theoretical basis for the application of fractional calculus to viscoelasticity. Journal of Rheology. 1983;**27**(3):201-210

[29] Atangana A, Alkahtani BST. New model of groundwater flowing within a confine aquifer: Application of Caputo-Fabrizio derivative. Arabian Journal of Geosciences. 2016;**9**(1):8

[30] Atangana A, Baleanu D. New fractional derivatives with nonlocal and non-singular kernel: Theory and application to heat transfer model. Journal of Thermal Sciences. 2015:1-8

[31] Sheikh NA, Ali F, Saqib M, Khan I, Jan SAA, Alshomrani AS, et al. Comparison and analysis of the Atangana–Baleanu and Caputo–Fabrizio fractional derivatives for generalized Casson fluid model with heat generation and chemical reaction. Results in Physics. 2017;**7**:789-800

[32] Sheikh NA, Ali F, Saqib M, Khan I, Jan SAA. A comparative study of Atangana-Baleanu and Caputo-Fabrizio fractional derivatives to the convective flow of a generalized Casson fluid. The European Physical Journal Plus. 2017;**132**(1):54

[33] Sheikh NA, Ali F, Khan I, Gohar M, Saqib M. On the applications of nanofluids to enhance the performance of solar collectors: A comparative analysis of Atangana-Baleanu and Caputo-Fabrizio fractional models. The European Physical Journal Plus. 2017;**132**(12):540

[34] Jan SAA, Ali F, Sheikh NA, Khan I, Saqib M, Gohar M. Engine oil based generalized brinkman-type nano-liquid with molybdenum disulphide nanoparticles of spherical shape: Atangana-Baleanu fractional model. Numerical Methods for Partial Differential Equations. 2017

[35] Alfvén H, Arrhenius G. Structure and evolutionary history of the solar system, I. Astrophysics and Space Science. 1970;**8**(3):338-421

[36] Seth GS, Kumbhakar B, Sarkar S. Unsteady MHD natural convection flow with exponentially accelerated free-stream past a vertical plate in the presence of hall current and rotation. Rendiconti del Circolo Matematico di Palermo. 2016;**1952**:1-21

Chapter 2

Recent Advances in Complex Fluids Modeling

Luís L. Ferrás, Maria L. Morgado, Magda Rebelo,
Rosalía T. Leiva, António Castelo,
Gareth H. McKinley and Alexandre M. Afonso

Additional information is available at the end of the chapter

http://dx.doi.org/10.5772/intechopen.82689

Abstract

In this chapter, we present a brief description of existing viscoelastic models, starting with the classical differential and integral models, and then focusing our attention on new models that take advantage of the enhanced properties of the Mittag-Leffler function (a generalization of the exponential function). The generalized models considered in this work are the fractional Kaye-Bernstein, Kearsley, Zapas (K-BKZ) integral model and the differential generalized exponential Phan-Thien and Tanner (PTT) model recently proposed by our research group. The integral model makes use of the relaxation function obtained from a step-strain applied to the fractional Maxwell model, and the differential model generalizes the familiar exponential Phan-Thien and Tanner constitutive equation by substituting the exponential function of the trace of the stress tensor by the Mittag-Leffler function. Since the differential model is based on local operators, it reduces the computational time needed to predict the flow behavior, and, it also allows a simpler description of complex fluids. Therefore, we explore the rheometric properties of this model and its ability (or limitations) in describing complex flows.

Keywords: Mittag-Leffler, viscoelastic, memory function, fractional calculus, rheology

1. Introduction

Since viscoelastic materials are abundant in nature and present in our daily lives (examples are paints, blood, polymers, biomaterials, etc.), it is important to study and understand viscoelastic behavior. Therefore, in this chapter, we further develop the modeling of viscoelasticity making use of fractional calculus tools.

We start this section with some basic concepts that are needed to derive and understand classical and fractional viscoelastic models. These are trivial concepts such as force, stress, viscosity, Hooke's law of elasticity and also Newton's law of viscosity. Later, we evolve to more complex concepts of viscoelasticity that involve the knowledge of fractional calculus, integral and differential models.

It is well known that a *force* is any interaction that when unopposed will change the motion of an object/body. *Stress* is an internal resistance provided by the body itself whenever it is under deformation. Stress is defined as the intensity of internal forces developed in the material. The intensity of any quantity is defined as the ratio of the quantity to the area on which it is acting, leading to: Average Stress = Force/Area. If we want to know the stress in one material point, then we must take the limit of the area to zero. A good example on how stress works is given by imagining a person lying on top of thin layer of ice. When the person is lying down on the ice, the force (weight) divided by the area of the surface of the person in contact with the ice is smaller, when compared to the case when someone is standing up (the weight is the same, but the area in contact with the ice is smaller). Therefore, eventually, the ice will break due to the high internal stresses when the person is standing. Finally, we refer to *elasticity* as the ability of a body to resist a distorting influence and to return to its original size and shape when that influence or force is removed. See for example **Figure 1** where three springs are stretched. If we remove the weights attached to the springs, the spring would ideally return to its initial/natural position.

Figure 1(b) also shows an experiment where we observe that the force (mass times gravity) applied to the spring (increasing weight) is proportional to the displacement. This is known as

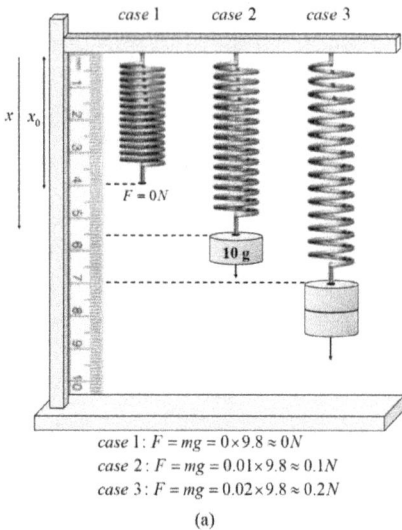

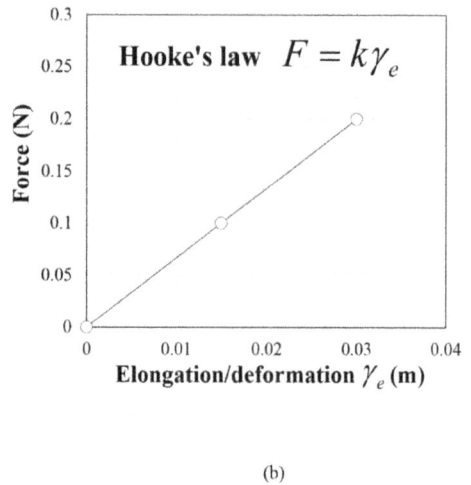

case 1: $F = mg = 0 \times 9.8 \approx 0 N$
case 2: $F = mg = 0.01 \times 9.8 \approx 0.1 N$
case 3: $F = mg = 0.02 \times 9.8 \approx 0.2 N$

(a)

(b)

Figure 1. Schematic of an experiment to obtain the relationship between force and deformation: (a) Experimental setup where three springs are stretched with the use of weights; (b) Graph showing the experimental results obtained from stretching three springs (the force is proportional to the deformation).

Hooke's law (the force F needed to extend or compress a spring by some distance $\gamma_e = (x - x_0)$ is proportional to that distance, $F = k\gamma_e$). Note that if we continue to increase the weight, eventually the spring will break. Therefore, Hooke's law is a very good linear approximation of what happens in the real world.

We will now explore the concept of *viscosity* in fluids. The viscosity of a fluid is a measure of the internal resistance to the rate of deformation:

As an example, imagine that we have a thin film of fluid in between two parallel plates, as shown in **Figure 2**. The fluid is at rest, and suddenly the upper plate starts moving with constant velocity U. This velocity will be felt at the bottom layer due to diffusion of momentum, and to keep the bottom wall fixed, we must exert a restraining force, that is measured with a force gage or dynamometer attached to that wall. Note that if we take the view of this portion of fluid as infinitesimally thin layers, we observe that each layer will drag the underlying layer due to the action of viscosity (internal resistance). The higher the viscosity, the more force will be required to deform the fluid at a given speed U.

Since the velocity of the thin layer adjacent to the top wall is U and the velocity of the bottom layer is 0, the velocity of each layer (for a Newtonian fluid) is given by $u(y) = Uy/h$, with y the coordinate shown in **Figure 2(a)**. **Figure 2(b)** shows the experimental forces measured for different ratios of U/h. We observe that the force is proportional to U/h and $U/h = du(y)/dy$; therefore, we conclude the following (*Newton's law of viscosity*):

$$\frac{Force}{Area} = \sigma = \eta\frac{U}{h} = \eta\frac{du(y)}{dy} \Rightarrow \sigma = \eta\frac{du(y)}{dy} \tag{1}$$

with σ the unidirectional stress and η as the constant of proportionality, known as the Newtonian shear viscosity. Note that du/dy is known as the rate of shear deformation, usually denoted by $\dot{\gamma}$.

A good example of something we may see every day and something that verifies Newton's law of viscosity is a dashpot. It is used for example as a door closer to prevent it from slamming shut.

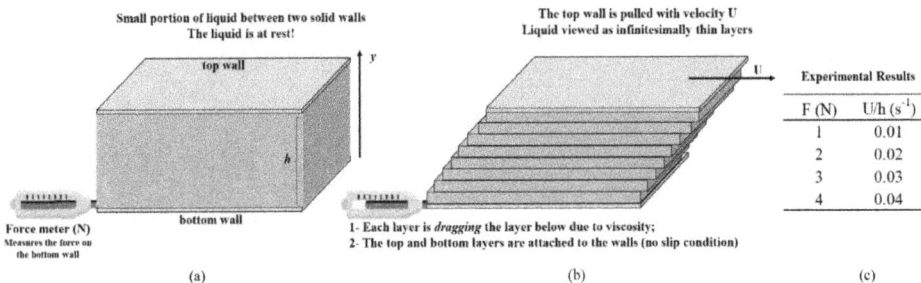

Experimental Results	
F (N)	U/h (s^{-1})
1	0.01
2	0.02
3	0.03
4	0.04

Figure 2. Schematic of an experiment to verify *Newton's law* of viscosity: (a) Liquid at rest between parallel plates; (b) The top wall is pulled with velocity U and a force meter is used to measure the force exerted on the bottom wall; (c) Experimental results.

1.1. Viscoelastic models

The simplest model that considers both viscous and elastic behavior is the linear Maxwell model [1] and can be obtained from a combination in series of a dashpot, $\sigma = \eta d\gamma_f(t)/dt$, and a spring, $\sigma = G\gamma_e(t)$ (with the subscripts f and e standing for Newtonian fluid and Hookean elastic solid, respectively), as shown in **Figure 3**.

The total deformation γ is the sum of the deformation obtained from the spring γ_e and the dashpot γ_f, and the rate of deformation is given by:

$$\frac{d\gamma(t)}{dt} = \frac{d\gamma_f(t)}{dt} + \frac{d\gamma_e(t)}{dt}$$
$$\Leftrightarrow \frac{d\gamma(t)}{dt} = \frac{\sigma}{\eta} + \frac{1}{G}\frac{d\sigma}{dt} \tag{2}$$
$$\Leftrightarrow \underbrace{\sigma + \lambda\frac{d\sigma}{dt} = \eta\frac{d\gamma(t)}{dt}}_{\text{Maxwell Model}}, \quad \lambda = \frac{\eta}{G}$$

The three-dimensional version of this model can be easily obtained by considering appropriate tensors instead of the scalar properties of stress and deformation, leading to the following model:

$$\boldsymbol{\sigma} + \lambda\frac{d\boldsymbol{\sigma}}{dt} = \eta\frac{d\boldsymbol{\gamma}(t)}{dt} \tag{3}$$

with $\boldsymbol{\sigma}$ the stress tensor, $\dot{\boldsymbol{\gamma}} = \left(\nabla\boldsymbol{u} + (\nabla\boldsymbol{u})^{\mathsf{T}}\right)$ the rate of deformation tensor, $\boldsymbol{u}$ the velocity vector, λ the relaxation time of the fluid and η the zero shear rate viscosity. This model can be equivalently written in integral form as

$$\boldsymbol{\sigma}(t) = \int_0^t Ge^{-(t-t')/\lambda}\frac{d\boldsymbol{\gamma}}{dt'}dt', \tag{4}$$

where $G = \eta/\lambda$ and it was assumed that the fluid is at rest for $t < 0$.

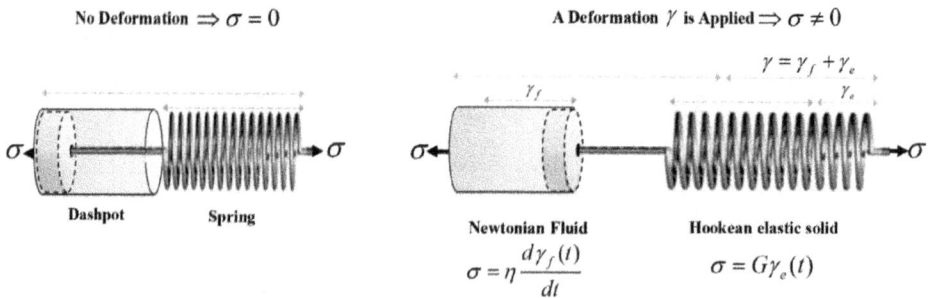

Figure 3. Maxwell model.

The Maxwell model is not *observer independent (frame invariant)* and, therefore, the results obtained with this model may not be correct if large deformations are considered (e.g., we may obtain a viscosity that depends directly on the velocity rather than the velocity gradient, which is not correct, and is unphysical). To solve this problem, new models were proposed in the literature that can deal with this non-invariance problem.

Two well-known examples of frame invariant models are the upper-convected Maxwell (UCM) model given by $\boldsymbol{\sigma} + \lambda\, \hat{\boldsymbol{\sigma}} = \eta\dot{\boldsymbol{\gamma}}$ (with $\hat{\boldsymbol{\sigma}} = \partial\boldsymbol{\sigma}/\partial t + \mathbf{u} \cdot \nabla\boldsymbol{\sigma} - (\nabla\mathbf{u})^T \cdot \boldsymbol{\sigma} - \boldsymbol{\sigma} \cdot \nabla\mathbf{u}$ the upper-convected derivative) that can also be written in integral form as

$$\boldsymbol{\sigma}(t) = \int_0^t \frac{\eta}{\lambda^2} e^{-(t-t')/\lambda}\left(\mathbf{C}_{t'}^{-1} - \mathbf{I}\right)dt' \tag{5}$$

where $\mathbf{C}^{-1}$ is the Finger strain tensor (a frame-invariant measure of deformation) [1]. The term $m(t - t') = \eta/\lambda^2 e^{-(t-t')/\lambda}$ is known as the memory function (the derivative of the relaxation modulus $G(t - t')$). Note that the relaxation modulus can be easily obtained by imposing a step strain (constant deformation), as shown in **Figure 4**, resulting in $G(t) = \sigma/\gamma_0 = G e^{-(t-t')/\lambda}$.

Other well-known example of a frame-invariant but now nonlinear viscoelastic model is the variation of the K-BKZ [2] model proposed by Wagner, Raible and Meissner [3, 4],

$$\boldsymbol{\sigma}(t) = \int_0^t m(t - t')h(I_1, I_2)\left(\mathbf{C}_{t'}^{-1} - \mathbf{I}\right)dt', \tag{6}$$

where $\mathbf{C}^{-1}$ is the Finger tensor [1], I_1, I_2 are the traces of $\mathbf{C}^{-1}$ and $\mathbf{C}$, respectively, and $h(I_1, I_2)$ is termed the *damping function* [5] (note that it is again assumed that the fluid is at rest for $t < 0$). A large number of damping functions can be found in the literature (see [5]). The term $m(t - t')$ was proposed to be of the form:

$$m(t - t') = \frac{a}{\lambda} e^{-(t-t')/\lambda}, \tag{7}$$

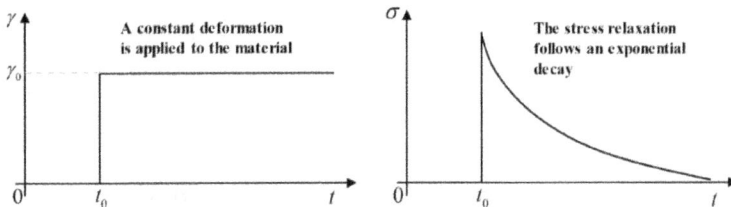

Figure 4. Step strain of a Maxwell model. The step strain is given by $\gamma = \gamma_0 H(t - t_0)$ with $H(t)$ the Heaviside function, and the stress relaxation is the solution of $\sigma + \lambda\, d\sigma/dt = \eta d\gamma_0 H(t - t_0)/dt$ with $\sigma(t_0) = \sigma_0$, given by $\sigma = \sigma_0 e^{-(t-t_0)/\lambda}$ ($\sigma_0 = G\gamma_0$).

where a and λ are model parameters. Note that the relaxation modulus is the response of the stress to a step in deformation (see **Figure 4**). It should be remarked that when $a = \eta/\lambda$ and $h(I_1, I_2) = 1$, we recover the integral version of the UCM model.

Different differential models were proposed in the literature along the years, with the aim of improving the modeling of complex viscoelastic materials, and with the aim of achieving the same modeling quality of integral models (by only using differential operators). Note that integral models are non-local (in time) operators that take into account all the past deformation of the fluid while differential models ones describe the material response in terms of the rate of change of stress to the *local deformation*, thus influencing the fitting quality of the model and the computational effort to numerically solve them (when performing numerical simulations).

More recently, new models have been proposed in the literature that basically take advantage of the generalization of the exponential function appearing in Eqs. (4), (5), and (7), thus allowing a more broad and accurate description of the relaxation of complex fluids (while the commonly used continuum approach describes the fluid as a whole, with only one relaxation, unless a Prony series is considered, that is, considering a series of the form $\sum_i a_i e^{-(t-t')/\lambda_i}$). This generalized function is the Mittag-Leffler function that naturally arises when solving problems involving fractional derivatives (more precisely, derivatives of non-integer order). This function will be introduced later in Section 3.

2. Fractional derivatives

To understand the need and the concept of a fractional derivative and its importance in the context of modeling physical processes, let us start with a simple example (**Figure 5**).

Imagine a portion of material that is principally formed of two different regions. In these regions, two similar physical processes φ_1 and φ_2 occur (for the time being it does not matter what is the process under study), but, at different rates, $d\varphi_1/dt = 0.1$ for Region I and $d\varphi_2/dt = 1$ for Region II. If we look at the portion of material as a whole, one would naturally choose the rate of 1 as representative of the material's behavior, because this region is bigger

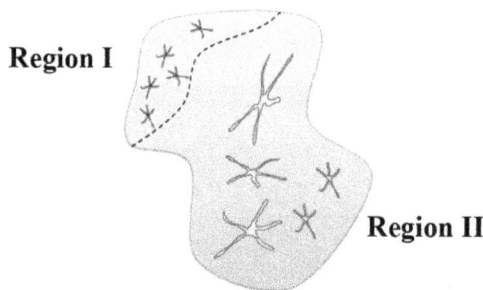

Figure 5. Material formed by two regions where the same physical process occurs at different rates.

(when compared to Region I). However, this clearly neglects entirely the local variation in the deformation associated with the neighboring Region I. With the help of fractional calculus, we may define derivatives/rates of non-integer order, and we may have (for example) a rate given by $d^\beta \varphi / dt^\beta$ with $\beta = 0.9$ (possibly better representing the material behavior as a whole, by providing intermediate rates).

Although we have not defined yet what a fractional derivative is, the fact of having the possibility of non-integer derivatives seems quite attractive, allowing the creation of a continuous path between integer-order derivatives that may lead to a better description of the different rates of a certain physical process occurring in the same material. This means that fractional derivatives can *transport* more and more precise local information from the microscopic world to the continuum description.

2.1. Riemann-Liouville and Caputo fractional derivatives

Now, to understand a fractional derivative, we start by acknowledging that the n-fold integral of a generic function $f(t)$ is given by the formula

$$\underbrace{\int_a^t \int_a^t \cdots \int_a^t f(t) dt dt \ldots dt}_{n\ times} = J_a^n f(t) = \frac{1}{(n-1)!} \int_a^t (t-t')^{n-1} f(t') dt'. \tag{8}$$

A generalization to non-integer values of n can be performed using the Euler Gamma function $\Gamma(x)$, leading to the Riemann-Liouville fractional integral

$$J_a^n f(t) = \frac{1}{\Gamma(\alpha)} \int_a^t (t-t')^{\alpha-1} f(t') dt', \tag{9}$$

where we have used α to represent the generalization of n to non-integer values. A fractional derivative of any order can then be obtained by manipulating the number of integrations and differentiations of the function $f(t)$. By performing the $m - \alpha$-fold integration of the mth derivative of $f(t)$, $J_a^{m-\alpha} D^m f(t)$ with $m = \lceil \alpha \rceil$, we arrive at the generalized derivative formula (Caputo fractional derivative [6]) of order $m - 1 < \alpha < m$,

$$\frac{d^\alpha f(t)}{dt^\alpha} = \frac{1}{\Gamma(m-\alpha)} \int_a^t (t-t')^{-\alpha+m-1} \frac{d^m f(t')}{dt'^m} dt', \quad m - 1 < \alpha < m, \tag{10}$$

This last fractional derivative is the one chosen to deal with physical processes due to the ease in handling initial and boundary conditions [7].

Next, we present two models that rely on the Mittag-Leffler function (a function closely related to fractional calculus) to improve their modeling and fitting capabilities when describing the behavior of viscoelastic materials. These are the fractional K-BKZ (integral) and the generalized Phan-Thien and Tanner (differential) models.

3. Viscoelastic models based on the Mittag-Leffler function

3.1. The fractional K-BKZ model

We first note that the Maxwell-Debye relaxation of stress (exponential decay—see Eqs. (4) and (5)) is quite common, but there are many real materials showing different types of fading memory, such as a power law decay $G(t) \sim t^{-\alpha}, 0 < \alpha < 1$ [8]. For example, the critical gel model investigated by Winter and Chambon is written $G(t) = St^{-\alpha}$. If we assume the relaxation modulus for an arbitrary loading history in such materials is given by $G(t - t') = \mathbb{V}(\Gamma(1 - \alpha))^{-1}(t - t')^{-\alpha}$ ($\mathbb{V}$ is known as a quasi-property [9] and is connected to the critical gel strength by $S = \mathbb{V}/\Gamma(1 - \alpha)$), then we have that:

$$\boldsymbol{\sigma}(t) = \frac{1}{\Gamma(1 - \alpha)} \int_0^t \mathbb{V}(t - t')^{-\alpha} \frac{d\gamma}{dt'} dt'. \tag{11}$$

By recognizing that the Caputo fractional derivative of a general function $\gamma(t)$ (in our case $\gamma(t)$ is the deformation) is defined as [10]:

$$\frac{d^\alpha \gamma(t)}{dt^\alpha} = \frac{1}{\Gamma(1 - \alpha)} \int_0^t (t - t')^{-\alpha} \frac{d\gamma}{dt'} dt', \tag{12}$$

we obtain a generalized viscoelastic model [10, 11], that can be written in the simple compact form:

$$\boldsymbol{\sigma} = \mathbb{V} \frac{d^\alpha \gamma(t)}{dt^\alpha}, \ 0 < \alpha < 1, \tag{13}$$

This model provides a generalized viscoelastic response, in the sense that when $\alpha = 1$ we obtain a Newtonian fluid, and when $\alpha = 0$ we obtain a Hookean elastic solid. The corresponding mechanical element is *intermediate* to the spring and dashpot shown in **Figure 3** and is thus known as a *spring-pot* [11, 12]. Note that care must be taken when $\alpha \to 1$ because of the singularity in $\Gamma(1 - \alpha)$ [12].

We can define the fractional Maxwell model (FMM) as a combination of two linear fractional elements (spring-pots) in series. In a series configuration, the stress felt by each spring-pot is the same, that is, $\boldsymbol{\sigma} = \mathbb{V} d^\alpha \gamma_1(t)/dt^\alpha = \mathbb{G} d^\beta \gamma_2(t)/dt^\beta, \ 0 < \alpha, \beta < 1$, and the total deformation is given by the sum of the deformation obtained for each spring-pot, $\gamma(t) = \gamma_1(t) + \gamma_2(t)$. The FMM can then be written as

$$\boldsymbol{\sigma}(t) + \frac{\mathbb{V}}{\mathbb{G}} \frac{d^{\alpha - \beta} \boldsymbol{\sigma}(t)}{dt^{\alpha - \beta}} = \mathbb{V} \frac{d^\alpha \gamma(t)}{dt^\alpha}, \tag{14}$$

This model allows a much better fit of rheological data, as shown in [12] but it is not frame invariant. However, following the same procedure employed with the Maxwell and K-BKZ model, that is, using the derivative of the relaxation function obtained for the Maxwell model as the memory function of the K-BKZ model, one can also use the derivative of the relaxation function of the FMM and insert it in the K-BKZ model, thus, obtaining a frame-invariant constitutive model, that retains all the good fitting properties of the FMM.

The relaxation function of the FMM can be obtained by solving the fractional differential Eq. (14) considering a constant deformation $\gamma = \gamma_0 H(t)$ ($H(t)$ is the Heaviside function) together with $\sigma(t_0) = \sigma_0$, leading to the relaxation modulus $G(t) = \sigma(t)/\gamma_0$ given by:

$$G(t) = \mathbb{G}t^{-\beta}E_{\alpha-\beta,1-\beta}\left(-\frac{\mathbb{G}}{\mathbb{V}}t^{\alpha-\beta}\right),$$
(15)

where $E_{a,b}(z)$ is the generalized Mittag-Leffler function [7],

$$E_{\alpha,\beta}(z) = \sum_{k=0}^{\infty}\frac{z^k}{\Gamma(\alpha k + \beta)},$$
(16)

and a characteristic measure of the relaxation spectrum described by the two spring-pots in series is $\lambda = (\mathbb{V}/\mathbb{G})^{1/(\alpha-\beta)}$.

This leads to the fractional K-BKZ model proposed by Jaishankar and Mckinley [12, 13], with $m(t - t')$ the memory function [2] in Eq. (6) now given by.

$$m(t - t') = \frac{dG(t - t')}{dt'} = -\mathbb{G}(t - t')^{-1-\beta}E_{\alpha-\beta,-\beta}\left(-\frac{\mathbb{G}}{\mathbb{V}}(t - t')^{\alpha-\beta}\right),$$
(17)

Note that here the relaxation modulus $G(t - t')$ is the one obtained for the FMM. Please see [11–13] for more details. It should be remarked that the Mittag-Leffler function was used in the past by Guy Berry to describe polymeric materials exhibiting Andrade creep [14].

The fractional K-BKZ model is therefore given by:

$$\boldsymbol{\sigma}(t) = -\mathbb{G}\int_0^t (t - t')^{-1-\beta}E_{\alpha-\beta,-\beta}\left(-\frac{\mathbb{G}}{\mathbb{V}}(t - t')^{\alpha-\beta}\right)h(I_1,I_2)(\mathbf{C}_{t'}^{-1} - \mathbf{I})dt',$$
(18)

and we need to ensure that the integral converges (see the Foundations of Linear Viscoelasticity by Coleman and Noll [15]). The main problem seems to be the term $(t - t')^{-1-\beta}$ that diverges as $t' \to t$, and $\int_0^t (t - t')^{-1-\beta}dt'$ diverges. Also, the term $E_{\alpha-\beta,-\beta}(\dots)h(I_1,I_2)$ is finite $\forall t', t' \le t$. Therefore, we must have $(\mathbf{C}_{t'}^{-1} - \mathbf{I}) = O((t - t')^m), m \ge 1$ as $t' \to t$ so that $(t - t')^{-1-\beta}$ $(\mathbf{C}_{t'}^{-1} - \mathbf{I}) = O((t - t')^n), n \le 1$ and therefore the integral converges.

It can be easily shown [1] that a Taylor series expansion of $\mathbf{C}_{t'}^{-1} - \mathbf{I}$ about $t' = t$ leads to.

$$(\mathbf{C}_{t'}^{-1} - \mathbf{I}) = -\sum_{k=1}^{\infty}\frac{(t' - t)^k}{k!}A_k(t),$$
(19)

with $A_k(t)$ the Rivlin-Ericksen tensors. Note that these tensors can be obtained directly from the velocity field without having to find the strain tensor [16]. We may therefore conclude that the integral is convergent, assuming a smooth velocity field is provided/obtained. Note that this does not mean that convergence problems will not arise during numerical calculations.

In Refs. [11, 12, 17], the beneficial fitting qualities of this constitutive model framework are discussed in detail. Here, we are interested in determining to what extent the properties of the Mittag-Leffler function can be used to improve the fitting quality of differential models, and this will be discussed in the next subsection.

3.2. Generalized Phan-Thien and Tanner model

The previous integral model given by Eq. (18) allows a good fit to experimental rheological data, in flows with defined kinematics where $\mathbf{C}^{-1}$ can be computed explicitly; but, it would be desirable to obtain also an improved frame-invariant differential model, that is easier to handle both mathematically and numerically, when compared to integral models, for solving complex flow problems with spatially varying kinematics. The model to be presented was recently proposed by our research group [18], and basically takes advantage of the flexible functional form of the Mittag-Leffler function by inserting this function into the already well-known Phan-Thien and Tanner (PTT) model, replacing the classical linear and exponential functions of the trace of the stress tensor.

The original exponential PTT model [19, 20] is given by.

$$\exp\left(\frac{\varepsilon\lambda}{\eta}\sigma_{kk}\right)\sigma + \lambda\overset{\square}{\sigma} = \eta\dot{\gamma}, \qquad (20)$$

with $\overset{\square}{\sigma} = \partial\sigma/\partial t + \mathbf{u}\cdot\nabla\sigma - (\nabla\mathbf{u})^T\cdot\sigma - \sigma\cdot\nabla\mathbf{u} + \xi(\mathbf{D}\cdot\sigma - \sigma\cdot\mathbf{D})$ being the Gordon-Schowalter derivative and σ_{kk} the trace of the stress tensor. Here, the parameter ξ accounts for slip between the molecular network and points in the continuous medium. The model was derived from a Lodge-Yamamoto type of network theory for polymeric fluids, in which the network junctions are not assumed to move strictly as points of the continuum but instead they are allowed a certain effective slip as well as a rate of destruction that depends on the state of stress in the network. Phan-Thien proposed that an exponential function form would be quite adequate to represent the rate of destruction of junctions and in [17] it was shown that the Mittag-Leffler function could improve the quality of model fits to real data by allowing different forms for the rates of destruction.

The model is then given by

$$\Gamma(\beta)E_{\alpha,\beta}\left(\frac{\varepsilon\lambda}{\eta}\sigma_{kk}\right)\sigma + \lambda\overset{\square}{\sigma} = \eta\dot{\gamma}, \qquad (21)$$

where the factor $\Gamma(\beta)$ is used to ensure that $\Gamma(\beta)E_{\alpha,\beta}(0) = 1$.

This new model can further improve the accuracy of the description of real data obtained with the original exponential function of the trace of the stress tensor, as shown in [18].

4. Parametric study of the GPTT model

We will now present a detailed parametric study on the influence of the new parameters α, β (arising from the Mittag-Leffler function) on the rheological behavior of the generalized exponential PTT model.

4.1. Steady-state shear flows

As shown in [18], the steady shear viscosity is given by $\eta(\dot{\gamma}) = \sigma_{xy}(\dot{\gamma})/\dot{\gamma}$ with

$$\sigma_{xy}(\dot{\gamma}) = \frac{\eta\dot{\gamma} - \sigma_{xx}Wi\xi}{\Gamma(\beta)E_{\alpha,\beta}\left(\frac{\varepsilon\lambda}{\eta}\left(\frac{2-2\xi}{2-\xi}\right)\sigma_{xx}\right)}, \tag{22}$$

and σ_{xx} is given by the solution of

$$\Gamma^2(\beta)E_{\alpha,\beta}\left(\frac{\varepsilon\lambda}{\eta}\left(\frac{2-2\xi}{2-\xi}\right)\sigma_{xx}\right)^2\sigma_{xx} = (2-\xi)(\lambda\dot{\gamma})^2\left[\frac{\eta}{\lambda} - \sigma_{xx}\xi\right]. \tag{23}$$

Here $Wi = \lambda\dot{\gamma}$ is the dimensionless strength of the shear flow and $\eta, \lambda, \varepsilon, \xi, \alpha, \beta$ are the constitutive parameters of the generalized PTT (or GPTT) model.

Since we consider a simple plane shear flow aligned with the x-axis, we have that $\sigma_{yy}(\dot{\gamma}) = \sigma_{xx}\xi/(2-\xi)$ and $\sigma_{zz}(\dot{\gamma}) = 0$ (see [18] for more details).

Eqs. (22) and (23) can readily be solved using the Newton-Raphson method (solving first Eq. (23) and then substituting the numerical values obtained for σ_{xx} into Eq. (22)).

Figure 6 shows the dimensionless steady shear viscosity obtained for the different parameters of the Mittag-Leffler function, α, β. On the left, we show the influence of α by keeping constant all other parameters. On the right, we show the influence of β (it should be remarked that when $\alpha, \beta = 1$ the exponential PTT model is recovered). We observe that when compared to the

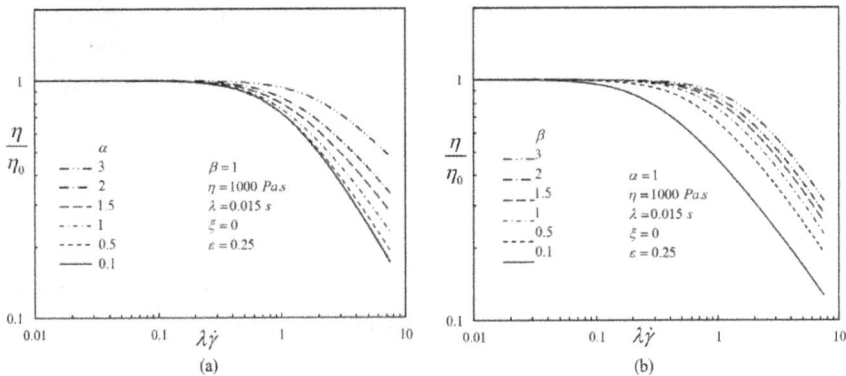

Figure 6. Dimensionless shear viscosity obtained for the different parameters of the Mittag-Leffler function; (a) varying α holding the other five parameters constant and (b) varying β.

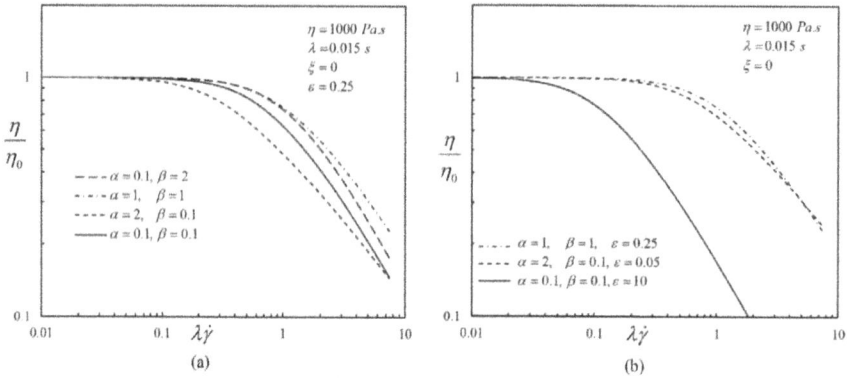

Figure 7. Dimensionless shear viscosity obtained for the different parameters of the Mittag-Leffler function varying: (a) Constant ε; (b) Varying ε.

classical exponential PTT model, when $\alpha, \beta < 1$, shear-thinning occurs for lower dimensionless shear rates and when $\alpha, \beta > 1$ there is a delay in the shear-thinning effect. For $\alpha, \beta > 1$ the shear viscosity increases, especially for high shear rates. Also, when we increase α, the slope of the shear viscosity curve for high dimensionless shear rates decreases (observed in **Figure 6(a)**), while varying β, the slope seems to be the same, but a higher viscosity is obtained (observed in **Figure 6(b)**).

Figure 7(a) shows the dimensionless steady shear viscosity, now obtained for different values of α, β and ε. These plots allow one to see that the ε parameter may not be compared directly to the value used in the classical models (featuring linear and exponential functions of the trace of the stress tensor). For comparison purposes, we plot again the curve obtained for the exponential PTT model with $\varepsilon = 0.25$ ($\alpha = \beta = 1$) by the dash-dot lines.

Note that (see **Figure 7(b)**) small variations of the parameter ε allows one to control the rate of transition to the shear-thinning at high Wi while maintaining a similar shear thinning set point.

Figure 7 shows that by setting different combinations of α, β we may obtain different slopes at higher dimensional shear rates. For low β and high α, we obtain a lower slope but a premature shear viscosity thinning, while for high β and low α, we obtain a higher slope but a delayed shear-thinning.

4.2. Steady-state elongational flows

The steady unidirectional extensional viscosity is defined as $\eta_E = (\sigma_{xx} - \sigma_{yy})/\dot{\varsigma}$, where $\dot{\varsigma}$ is the imposed elongation rate [18], and can be obtained by solving the system of equations (for a simpler technique that does not involve an iterative procedure, please consult [18])

$$\sigma_{xx}\left(\Gamma(\beta)E_{\alpha,\beta}\left(\frac{\varepsilon\lambda}{\eta}\sigma_{kk}\right) - 2\lambda\dot{\varsigma}(1-\xi)\right) = 2\eta\dot{\varsigma}, \qquad (24)$$

$$\sigma_{xx}\left(\Gamma(\beta)E_{\alpha,\beta}\left(\frac{\varepsilon\lambda}{\eta}\sigma_{kk}\right) + \lambda\dot{\varsigma}(1-\xi)\right) = -\eta\dot{\varsigma}, \tag{25}$$

with $\sigma_{kk} = \sigma_{xx} + 2\sigma_{yy}$.

Figure 8 shows the dimensionless steady elongational viscosity obtained for different parameters of the Mittag-Leffler function. In **Figure 8(a)**, we show the influence of α by keeping constant all other parameters. In **Figure 8(b)**, we show the influence of β. Note that we have used the same parameters as in the shear viscosity case.

Note that when we increase α, β, we observe an increase of the elongational viscosity, with the maximum value being reached for higher dimensionless extensional rates. Again, we observe different asymptotic slopes for high extension rates (when varying α). Note that there is no overshoot for low values of β.

We may conclude that by varying α, β, we change both the shear and elongational viscosities, and therefore the fit to experimental data should be performed with care, taking into account this dependence.

Figure 9 shows the effect of the parameters used in **Figure 7**, for the case of elongational viscosity. The results are qualitatively similar to the ones obtained in **Figure 7**, that is, in terms of changes to the asymptotic slopes at high deformation rates and premature/delayed thinning. It can be observed that the elongational viscosity is more sensitive to changes in the parameters α, β and ε. This result is to be expected since this is a strong flow, and, the exponential PTT model was originally proposed to be able to describe the response of complex fluids in strong flows. **Figure 9(a)** shows that the overshoot can be suppressed using a low β and high α values. Note that the maximum extensional viscosity is obtained for the exponential PTT model, and that the values of α, β have a strong influence on the asymptotic slope of the curve for high extensional rates. **Figure 9(b)** shows three different curves for different combinations of α, β and ε. Note that for $\alpha = 0.1, \beta = 0.1$ and $\varepsilon = 10$ we can also suppress the

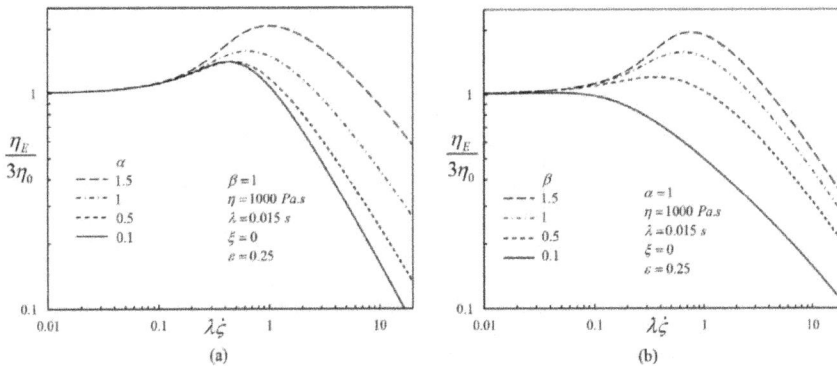

Figure 8. Dimensionless elongational viscosity obtained for different parameters of the Mittag-Leffler function: (a) Varying α; (b) Varying β.

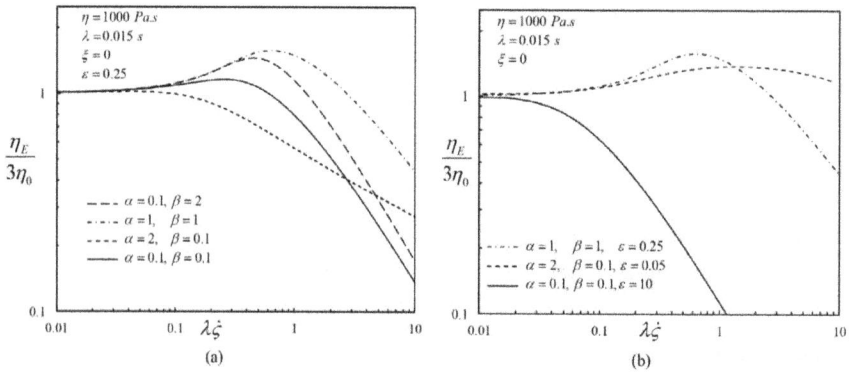

Figure 9. Dimensionless elongational viscosity obtained for the different parameters of the Mittag-Leffler function: (a) Constant ε; (b) Varying ε.

overshoot in the extensional viscosity, and for $\alpha = 2, \beta = 0.1$ and $\varepsilon = 0.05$ we can decrease the curvature of the overshoot, and at the same time decrease the slope of curve.

4.3. Steady-state shear and elongational flows

Until now, we have explored generally the influence of the different model parameters on the behavior of the GPTT model for steady flows, but, a more quantitative side-by-side comparison

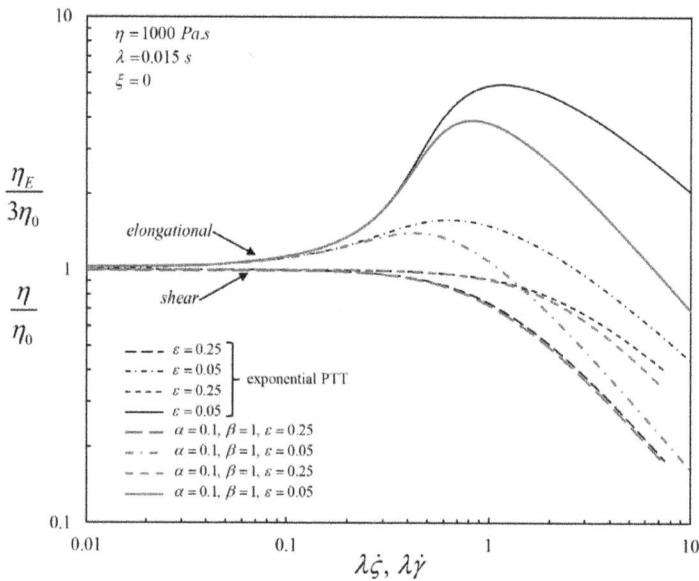

Figure 10. Comparison of the dimensionless elongational and shear viscosity obtained for different parameters of the Mittag-Leffler function, varying ε, and the classical exponential PTT model ($\alpha = 1, \beta = 0$).

between the shear and elongational flow curves was not performed, and the limited flexibility of the classical exponential PTT model for fitting experimental data (when compared to the GPTT) was not explored. In **Figure 10**, we try to illustrate the advantages of using the Mittag-Leffler function instead of the classical exponential one. To this end, we present the viscometric predictions obtained for both shear and elongational flows for both models (GPTT and exponential PTT).

Figure 10 illustrates the additional flexibility of using the Mittag-Leffler function, by showing that we can manipulate the magnitude of the increase in the elongational viscosity and at the same time only slightly change the shear viscosity. This allows better fits to rheological data when using the Mittag-Leffler function [18]. Note that in the exponential PTT model, when we increase the ε parameter, both the shear and elongational viscosities increase concomitantly.

5. Conclusions

In this chapter, we have presented a brief introduction to the world of viscoelastic models capable of describing the rheology of complex fluids, and we have summarized some of the well-known classical differential and integral models.

With incorporation of ideas from fractional calculus, most of these models can be further improved, either by changing classical local operators for improved (non-local) fractional versions, or, either using new analytic functions that arise in the realm of fractional differential equations, such as the Mittag-Leffler function.

As an example, we present the fractional K-BKZ model and the recently proposed generalized PTT model. The fractional K-BKZ model allows a better description of fluid flow behavior (when compared to the generalized PTT model), but, increases the need for high computational power. Therefore, the novelty of the present work is our detailed study on the influence of the Mittag-Leffler function in shear and elongational flows of a generalized PTT model.

Acknowledgements

L.L. Ferrás and A.M. Afonso acknowledge the Project PTDC/EMS-ENE/3362/2014-POCI-010145-FEDER-016665—funded by FEDER funds through COMPETE2020—Programa Operacional Competitividade e Internacionalização (POCI) and by national funds through FCT—Fundação para a Ciência e a Tecnologia; I.P. L.L. Ferrás would also like to thank the funding by FCT through the scholarship SFRH/BPD/100353/2014. M.L. Morgado would like to thank the funding by FCT through Project UID/MULTI/04621/2013 and M. Rebelo would also like to thank the funding by FCT through Project UID/MAT/00297/2013 (Centro de Matemática e Aplicações).

Author details

Luís L. Ferrás[1], Maria L. Morgado[2,3], Magda Rebelo[4], Rosalía T. Leiva[5], António Castelo[5], Gareth H. McKinley[6] and Alexandre M. Afonso[7]*

*Address all correspondence to: aafonso@fe.up.pt

1 Department of Mathematics, University of Minho (CMAT-UM), Guimarães, Portugal

2 CEMAT, Instituto Superior Técnico, Universidade de Lisboa, Lisboa, Portugal

3 Departamento de Matemática, Universidade de Trás-os-Montes e Alto Douro, UTAD, Vila Real, Portugal

4 Center of Mathematics and Applications (CMA) and Departamento de Matemática, Faculdade de Ciências e Tecnologia, Universidade NOVA de Lisboa, Caparica, Portugal

5 Departamento de Matemática Aplicada e Estatística, Universidade de São Paulo, Brazil

6 Department of Mechanical Engineering, Massachusetts Institute of Technology, Cambridge, MA, USA

7 Centro de Estudos de Fenómenos de Transporte, Departamento de Engenharia Mecânica, Faculdade de Engenharia da Universidade do Porto, Porto, Portugal

References

[1] Bird RB, Armstrong RC, Hassager O. Dynamics of Polymeric Liquids Fluid Mechanics. 2nd ed. Vol 1. New York: Wiley; 1987. 672p. ISBN: 978-0-471-80245-7

[2] Bernstein B, Kearsley E, Zapas L. A study of stress relaxation with finite strain. Transactions of The Society of Rheology. 1963;7:391-410. DOI: 10.1122/1.548963

[3] Wagner MH. Analysis of stress-growth data for simple extension of a low-density branched polyethylene melt. Rheologica Acta. 1976;15:136-142. DOI: 10.1007/BF01517504

[4] Wagner MH, Raible T, Meissner J. Tensile stress overshoot in uniaxial extension of a LDPE melt. Rheologica Acta. 1979;18:427-428. ISSN: 0035-4511/ASTM-Coden: RHEAAK

[5] Joseph DD. Luis. International Symposium on Viscoelastic Fluids; Tobago, West Indies; 1994

[6] Caputo M. Linear models of dissipation whose Q is almost frequency independent-II. Geophysical Journal International. 1967;13:529-539. DOI: 10.1111/j.1365-246X.1967.tb02303.x

[7] Podlubny I. Fractional Differential Equations: An Introduction to Fractional Derivatives, Fractional Differential Equations, to Methods of Their Solution and Some of Their Applications. 1st ed. San Diego, California, USA: Springer; 1998. 340p. ISBN: 9780125588409

[8] Ng TS-K, McKinley GH, Padmanabhan M. Linear to non-linear rheology of wheat flour dough. Applied Rheology. 2006;16:265-274

[9] Ferrás LL, Ford N, Morgado L, Rebelo M, McKinley GH, Nóbrega JM. A primer on experimental and computational rheology with fractional viscoelastic constitutive models. AIP Conference Proceedings. 2017;**1843**:020002 . 1-13. DOI: 10.1063/1.4982977

[10] Schiessel H, Metzler R, Blumen A, Nonnenmacher T. Generalized viscoelastic models: Their fractional equations with solutions. Journal of Physics A: Mathematical and General. 1995;**28**:6567-6584. DOI: 10.1088/0305-4470/28/23/012

[11] Friedrich C. Relaxation and retardation functions of the Maxwell model with fractional derivatives. Rheologica Acta. 1991;**30**:151-158. DOI: 10.1007/BF01134604

[12] Jaishankar A, McKinley GH. A fractional K-BKZ constitutive formulation for describing the nonlinear rheology of multiscale complex fluids. Journal of Rheology. 2014;**58**: 1751-1788. DOI: 10.1122/1.4892114

[13] Freed A, Diethelm K. Fractional calculus in biomechanics: A 3D viscoelastic model using regularized fractional derivative kernels with application to the human calcaneal fat pad. Biomechanics and Modeling in Mechanobiology. 2006;**5**:203-215. DOI: 10.1007/s10237-005-0011-0

[14] Berry GC. The stress-strain behavior of materials exhibiting Andrade creep. Polymer Engineering and Science. 1976;**16**:777-781

[15] Coleman B, Noll W. Foundations of linear viscoelasticity. Reviews of Modern Physics. 1961;**33**:239-249. DOI: 10.1103/RevModPhys.33.239

[16] Phan-Thien N. Understanding Viscoelasticity. 1st ed. Berlin, Germany: Springer; 2002. 144p. ISBN: 3-540-43395-3

[17] Ferrás LL, Ford N, Morgado L, Rebelo M, McKinley GH, Nóbrega JM. Theoretical and numerical analysis of unsteady fractional viscoelastic flows in simple geometries. Computer and Fluids. 2018;**174**:14-33

[18] Ferrás LL, Morgado L, Rebelo M, McKinley GH, Afonso A. A generalised Phan-Thien-Tanner model. Journal of Non-Newtonian Fluid Mechanics. Submitted

[19] Phan-Thien N, Tanner RI. New constitutive equation derived from network theory. Journal of Non-Newtonian Fluid Mechanics. 1977;**2**:353-365. DOI: 10.1016/0377-0257(77)80021-9

[20] Phan-Thien N. A nonlinear network viscoelastic model. Journal of Rheology. 1978;**22**: 259-283. DOI: 10.1122/1.549481

Chapter 3

Super-Speeding Jets in MHD Couette Flow

Krzysztof Mizerski

Additional information is available at the end of the chapter

http://dx.doi.org/10.5772/intechopen.79005

Abstract

A magnetohydrodynamic flow of a viscous and conducting fluid confined between two parallel differentially moving boundaries is considered. The whole system is in a *strong* magnetic field chosen in such a way that the Hartmann boundary layers which form in this problem become singular at the points where the magnetic field becomes tangent to the boundary. Two geometries are taken into account: *plane* and *spherical*. Within the class of such configurations, the velocity field of the fluid and the influence of the conductivity of the boundaries on the fluid's motion are reviewed here. In the region of singularity, where the magnetic field is tangent to the boundary, the fluid's velocity exceeds that of the moving boundary. The effect of nonzero conductivity of the boundaries on the super-speeding jets is vital and has been enlightened in a series of papers, including experimental and theoretical findings. The mechanism of the formation of super-speeding jets in the considered configurations has been explained, which is based on strong Hartmann currents allowed to enter the boundary layer due to the singularity. In the case of both perfectly conducting boundaries, the super velocity was shown to be as strong as to scale with the Hartmann number as $\mathcal{O}\left(M^{1/2}\right)$.

Keywords: super rotation, magnetohydrodynamics, MHD boundary layers, Hartmann layer singularity, nonzero conductivity

1. Introduction

Super-speeding jets in the geometry of magnetohydrodynamic (MHD) spherical Couette flow have been first noticed in the numerical simulations of Dormy et al. [1]. They have analyzed a flow of an electrically conducting fluid in a spherical gap between concentric spherical shells, rapidly and differentially rotating about a common axis in a centered axial dipolar magnetic field. The solid inner sphere, which had the same conductivity as the fluid, was spinning slightly faster than the insulating outer shell. The stationary flow obtained via DNS exhibited a super-rotating structure near the region, where the critical magnetic field line, henceforth denoted by $\mathcal{C}$, was tangent to the outer shell. The angular velocity of the flow in that region was about 50% greater than that of the inner sphere, which was driving the flow.

Hollerbach [2] for the first time investigated numerically the effect of nonzero conductivity of the outer shell in the same spherical geometry but the outer boundary was held motionless, thus eliminating the Coriolis force from the problem. He reported that the super rotation in the singular region was greatly enhanced and scaled with the value of the Hartmann number M.[1] Hollerbach [3] studied the MHD spherical Couette flow for several different topologies of the external field lines and also observed that singular points of isolated contact of the magnetic lines with boundaries result in the formation of jets.

In a following sequence of three theoretical papers, the mechanism of super-velocity formation and the effect of nonzero conductivity of the boundaries have been explained. Dormy et al. [4] performed a joint analytical and numerical study of the system analyzed previously by Hollerbach [2], where they have described the super rotating shear layer along the critical magnetic line $\mathcal{C}$ which grazes the outer boundary and found analytic expressions for super rotation within the scope of asymptotic theory for $M \gg 1$, confirmed by the results of numerical simulations. Not only they have explained the physics of the mechanism behind the formation of super-speeding jets in the studied configuration, which relies on the enhancement of the Lorentz force accelerating the flow, due to strong currents entering the singular Hartmann boundary layer at the outer shell near the point of contact of the critical field line $\mathcal{C}$ with the boundary, but also their analysis set grounds for the following theoretical findings. The study of Mizerski and Bajer [5] greatly relied on that of Dormy et al. [4], although it involved geometries—planar and spherical, and the resting boundary was weakly conducting (as opposed to the previous study, where it was insulating). The two geometries studied by Mizerski and Bajer [5] are depicted on **Figure 1**. In the plane geometry, the bottom boundary is moving at a constant speed and has the same conductivity as the fluid, whereas the conductivity of the upper boundary relative to the fluid's conductivity ϵ is assumed small at the order $\epsilon \sim M^{-1}$. The rest of the space was an insulator. The crucial features of the external field in this configuration are that it is potential and that there exists a critical magnetic line, which grazes the upper boundary thus creating a singularity of the Hartmann layer. The two configurations depicted on **Figure 1** are planar and spherical counterparts, equivalent from the point of view of physics of the super-speeding jets which form near the singularities. The plane configuration, however, captures all the necessary physical ingredients of the problem but at the same time makes the problem more transparent, avoiding the complications resulting from the curvature of the boundaries. Mizerski and Bajer [5] utilized this simplification and demonstrated the super-velocity excess, that is, the difference between the super velocity in the cases of a conducting and insulating outer boundary scales like $\mathcal{O}\left(\epsilon M^{3/4}\right)$. They have also studied a somewhat similar case of a strongly conducting, but thin outer shell, with conductivity being the same as that of the fluid, but the relative thickness of the conducting shell to the thickness

---

[1]He fitted an exponent of $M^{0.6}$ to his numerical results, which however, was later shown not to be the true asymptotic scaling law.

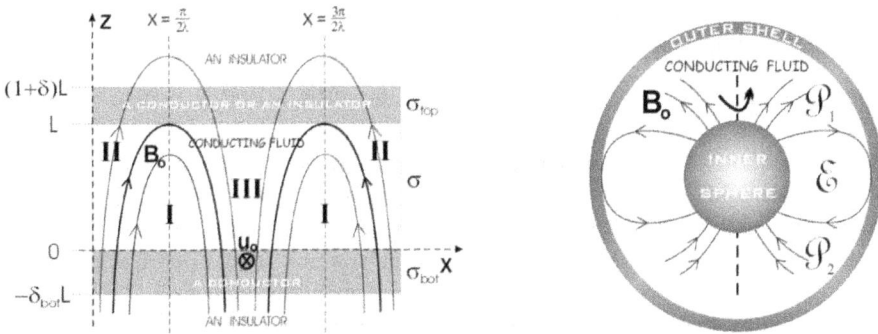

Figure 1. A sketch of the two situations considered: the plane case on the left (the bottom boundary is moving) and the spherical case on the right (the inner sphere is rotating). After Mizerski and Bajer [5].

of the fluid layer δ was assumed small, at the order $\delta \sim M^{-1}$. They demonstrated that both cases, $\epsilon \sim M^1$, $\delta \sim 1$ and the other one $\epsilon \sim 1$ and $\delta \sim M^{-1}$, are exactly equivalent in terms of the flow structure and the super-velocity magnitude. In the latter case, the super-velocity excess was shown to scale like $\mathcal{O}\left(\delta M^{3/4}\right)$. The same scalings were shown to pertain to the spherical geometry. The most notable contribution to the problem of super-speeding jets was the remarkable comprehensive analysis of Soward and Dormy [6] in the spherical geometry. They have emphasized the role of the parameter $\epsilon M^{3/4}$ (or $\delta M^{3/4}$ for the case of thin outer shell), identified in Mizerski and Bajer [5], for the general case of continuously varying relative conductivity of the outer shell ϵ from zero to infinity. They have reported the following scaling laws for the super velocity (angular velocity) in the singular region, for the case of perfectly conducting inner boundary.

$$\Omega_{max} = \mathcal{O}\left(M^{1/2}\right) \quad \text{for} \quad 1 \ll \epsilon \ll \epsilon M^{3/4}$$

$$\Omega_{max} = \mathcal{O}\left(\varepsilon^{2/3} M^{1/2}\right) \quad \text{for} \quad \epsilon \ll 1 \ll \epsilon M^{3/4} \tag{1}$$

$$\Omega_{max} = \mathcal{O}(1) \quad \text{for} \quad \epsilon \ll \epsilon M^{3/4} \ll 1$$

The magnitude of the super rotation Ω_{max} was shown to be proportional to the magnitude of current on the critical $\mathcal{C}$-line, denoted by $\mathcal{J}_c$, i.e., $\Omega_{max} \sim \mathcal{J}_c$.

The phenomenon of super rotation was also observed in the experimental setup called "Derviche Tourneur Sodium" (DTS) located in Grenoble at the Université Joseph-Fourier. Nataf et al. [7] conducted experiments on the spherical Couette flow of liquid sodium in an external, centered axial dipolar field, with both boundaries differentially rotating. The outer shell was only 5 mm thick, about 27 times thinner than the fluid gap and about 8 times less electrically conductive than liquid sodium. The Hartmann number in the experiment was at the order of a thousand. They observed the super-rotating jets and obtained a very good agreement with the numerical models. However, they also observed that the super-speeding jets can be destabilized and reported oscillatory motion near the singular region. More recently, Brito et al. [8] further exploited the same DTS experimental setup and explored the effects of strong inertia. They also reported strong super rotation; however, they clearly demonstrated that the Coriolis force tends to suppress the super-speeding jets.

Wei and Hollerbach [9] investigated numerically the effect of strong inertia, that is, large Reynolds number, on the spherical Couette flow configuration with the outer shell stationary. Three configurations of the external magnetic field were chosen, which resulted from a combination of dipolar and axial fields. The super-speeding jets have been destabilized by increasing the Reynolds number, whereas strengthening the filed had the opposite effect. Most recently, Hollerbach and Hulot [10] performed numerical analysis of a similar problem in cylindrical geometry, putting an emphasis on the role of conductivity of the boundaries. The field configurations were also chosen so as to create singularities in the flow. When the boundaries were electrically conducting, super-speeding jets were reported on the contrary to the case with insulating boundaries, when simply shear layers were observed in the singular regions. A curious observation is made by the introduction of a nonzero azimuthal component of the external field in which case the conductivity of the boundaries has the opposite effect to the previous case, greatly suppressing the magnitude of super rotation.

The motivation for some of the aforementioned studies was justified on geophysical grounds. The investigations of the Earth's interior reveal differential rotation of the inner core (cf. [11, 12]) and that the electrical conductivity of the lower mantle is nonnegligible [13]. Moreover, some evidence can be found for the existence of a very thin layer of anomalously high conductivity at the base of the mantle [14, 15]. It must be said, however, that the model of MHD spherical Couette flow is so idealized with respect to the true dynamics of the core, neglecting thermal and compositional driving, turbulence, the solidification processes at the inner core, etc., that no direct comparisons with the flow at the core-mantle boundary can be made. Nevertheless, it might be possible that the effect of super rotation manifests itself on the field zero isolines locally at the core mantle boundary.

1.1. Ferraro's law of isorotation

Throughout this chapter, we will assume that the Hartmann number,

$$M = \frac{B_0 L}{\sqrt{\mu_0 \rho \nu \eta}} = \sqrt{\frac{\sigma}{\rho \nu}} B_0 L \gg 1 \tag{2}$$

is large. In the above, μ_0, ρ, ν, η, and σ are the magnetic permeability, constant density, viscosity, magnetic diffusivity, and electrical conductivity of the fluid, respectively; B_0 is the typical strength of the external magnetic field; and L is the distance between the boundaries.

In such a case, the Ferraro's law of isorotation states that for a steady azimuthal motion about an axis of symmetry of an electrically conducting fluid, the magnitude of the angular velocity is predominantly constant along a magnetic field line. This means that in the studied configurations presented in **Figure 1**, the flow in the equatorial region $\mathscr{E}$ in the spherical case and region **I** in the planar case both bounded by the critical line which grazes the outer/upper boundary must significantly differ from the flow outside those regions. The magnetic lines

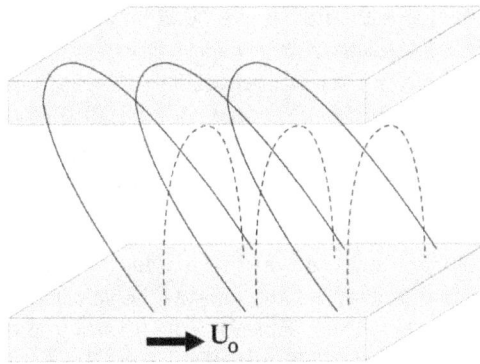

Figure 2. The moving electrically conducting boundary drags the field lines with it, but only those lines which experience drag from the top, stationary boundary are tilted (solid lines). The lines within the arcade bounded by the critical C-line are carried with the same velocity as that of the bottom boundary. After Mizerski and Bajer [5].

within the regions $\mathscr{E}$ and $\mathbf{I}$ do not reach the outer/upper boundary and their both footpoints lie on the inner/bottom boundary, which is moving. Since the moving boundary is assumed electrically conducting with the same conductivity as the fluid, the magnetic field lines within the arcade bounded by the C-line are carried by the fluid without any tilt. Therefore, by the Ferraro's law, the flow within the arcade must be uniform, with the same magnitude as the velocity of the moving boundary. On the contrary, the magnetic lines outside those regions extend from one boundary to the other; therefore, they are tilted due to advection by the inner/bottom boundary and the drag they experience from the outer/top, stationary boundary. The effect of competition of the moving and the stationary boundary makes the flow vary from one field line to the other. This is illustrated in **Figure 2** for the case of planar geometry.

In the following, we review the analytic approach and most important results for the two cases introduced in **Figure 1**.

2. Mathematical formulation

We study two types of stationary, magnetohydrodynamic Couette flow, that is, a flow between two parallel boundaries one of which is moving with a constant velocity: plane and spherical. The flow interacts with a strong (large Hartmann numbers) force-free magnetic field tangent to the boundaries at some isolated points. In the spherical case, the external field is a dipole field with a source at the center of the system and in the plane case, it is harmonic with oscillatory dependence in the direction perpendicular to the velocity of the moving boundary with an arbitrary period $\frac{2\pi}{k}$. **Figure 1** illustrates both the spherical and the plane cases.

We focus here on the phenomenon of super velocities in the regions of singularity of the Hartmann boundary layers which are present in this problem, that is, in the vicinity of points, where the magnetic field becomes tangent to the stationary boundary. In those regions, the fluid's velocity exceeds the velocity of the moving boundary. The aim of this chapter is to review the influence of conductivity of the upper/outer boundary on the enhancement of the super-velocity magnitude and explain why the super velocities are larger in the case when the stationary boundary is conducting when compared to the case where it is insulating. As mentioned in the introduction, this fact was proved numerically by several authors. We adopt here the analytic approach and notations of Dormy et al. [4] and Mizerski and Bajer [5]. Majority of the analysis will be done in the simpler and therefore more transparent planar geometry.

We consider here a stationary state in which the velocity of the fluid and the induced magnetic field have only one component, the same as the velocity of the moving boundary, axisymmetric for the spherical case and translationally invariant in the direction of the flow for the flat case. Small differential rotation/motion of the boundaries is assumed for the Couette flow dominated by the magnetic forces, that is, the magnetic Reynolds number is assumed small,

$$Re_M = \frac{Lu_0}{\eta} \ll 1 \tag{3}$$

where u_0 is the velocity of the moving boundary and the Hartmann number (2) is large. The above assumption of small Re_M implies that the flow-induced component of the magnetic field is of the order Re_M, and thus the magnetic field is decomposed in the following way

$$\mathbf{B} = \mathbf{B_0}(x,z) + Re_M b(x,z)\widehat{\mathbf{e}}_y \quad \text{for the plannar case} \tag{4}$$

$$\mathbf{B} = \mathbf{B_0}(s,z) + Re_M b(s,z)\widehat{\mathbf{e}}_\varphi \quad \text{for the spherical case} \tag{5}$$

where (s, φ, z) are the cylindrical polar coordinates, $\mathbf{B_0}$ is the external potential field, and $Re_M \mathbf{b}$ is the perturbation magnetic field generated by the flow. Indeed, in the numerical simulations of Hollerbach and Skinner [16] for infinitesimally small rotation rate, the flow was axisymmetric with only the azimuthal components of the velocity and the induced magnetic field present. The assumption of the small magnetic Reynolds number is crucial for the spherical case to neglect the nonlinear term which does not vanish because of the curvature effects. In the flat case, however, this assumption is not necessary to simplify the equations, because the nonlinear term vanishes due to the translational symmetry in the direction of the flow. Nevertheless, we keep the Reynolds numbers small even in the plane flow, since for high Rm, the unidirectional solutions are most probably unstable.

Furthermore, the solution for the plane flow is also valid when both boundaries are moving with different velocities since it is just a matter of changing the frame of reference to one moving at the same constant velocity as one of the boundaries. In the spherical case, however, when both boundaries rotate at different angular velocities, the Coriolis force substantially modifies the solution even in the case of small differential rotation unless the flow is strongly dominated by the magnetic force. The problem of MHD Couette flow with Coriolis force was investigated numerically by Hollerbach [17] and Dormy et al. [1] and analytically, for small Elsasser numbers, by N. Kleeorin et al. [18]. As remarked in the introduction, Brito et al. [8] demonstrated experimentally the detrimental effect of the Coriolis force on superrotation.

2.1. The equations and the main flow solution

As mentioned, we present the analysis for the flat case illustrated on the left panel of **Figure 1**. In Cartesian coordinates (x, y, z), the lower boundary is moving in the "y" direction and the "z" axis is perpendicular to both parallel boundaries. The dimensionless external magnetic field is given by

$$\mathbf{B_0}(x, z) = \nabla A \times \hat{\mathbf{e}}_y = \nabla \Phi = \left[e^{-\lambda z} \sin \lambda x, 0, e^{-\lambda z} \cos \lambda x \right], \tag{6}$$

where $A = \exp(-\lambda z) \sin \lambda x / \lambda$, $\Phi = -\exp(-\lambda z) \cos \lambda x / \lambda$, and $2\pi/\lambda$ is the arbitrary period of oscillation of the external field in the "x" direction.

The lower moving boundary is assumed to have the same conductivity as the fluid, while the conductivity of the upper one, which is at rest,

$$\epsilon = \frac{\sigma_u}{\sigma_f} \tag{7}$$

can vary from zero to infinity, where σ_u and σ_f are the electrical conductivities of the upper boundary and the fluid, respectively. The magnetic permeabilities of boundaries and the fluid are assumed to be same.

The general set of equations for the analyzed stationary state is obtained by taking the "y" components of the induction and the Navier-Stokes equations

$$\begin{aligned} \mathbf{B_0} \cdot \nabla u + \nabla^2 b &= 0 \\ \mathbf{B_0} \cdot \nabla b + \frac{1}{M^2} \nabla^2 u &= 0 \end{aligned} \quad \text{for} \quad 0 < z < 1 \tag{8}$$

with $M \gg 1$ and no-slip boundary conditions for the velocity field $u(x, z)$

$$\begin{aligned} u(x, 1) &= 0 \\ u(x, 0) &= 1. \end{aligned} \tag{9}$$

Inside the rigid conductors, the magnetic field $\mathbf{b}$ has to satisfy Laplace equations:

$$\begin{aligned} \nabla^2 b_1 &= 0 \quad \text{for} \quad 1 < z < 1 + \delta_1 \\ \nabla^2 b_2 &= 0 \quad \text{for} \quad -\delta_2 < z < 0 \end{aligned} \tag{10}$$

where b_i and δ_i are the perturbation magnetic fields inside the rigid conductors and their dimensionless thickness (with $i = 1$ for the upper conductor and $i = 2$ for the lower one). At the boundaries with the insulator at $z = 1 + \delta_1$ and at $z = -\delta_2$, the perturbation magnetic field must vanish since there is no imposed magnetic field in the "y" direction

$$\begin{aligned} b_1(x, 1 + \delta_1) &= 0 \\ b_2(x, -\delta_2) &= 0 \end{aligned} \tag{11}$$

Finally, the conditions at $z = 0$ and at $z = 1$ for the magnetic field can be written in the following form

$$\begin{cases} b(x,0) = b_2(x,0) \\ \left.\frac{\partial b}{\partial z}\right|_{z=0} = \left.\frac{\partial b_2}{\partial z}\right|_{z=0} \end{cases} \text{ and } \begin{cases} b(x,1) = b_1(x,1) \\ \left.\epsilon\frac{\partial b}{\partial z}\right|_{z=1} = \left.\frac{\partial b_1}{\partial z}\right|_{z=1} \end{cases} \tag{12}$$

To understand the structure of the flow, it is very important to note the symmetries in the system with respect to planes defined by $x = n\frac{\pi}{2\lambda}$ for $n \in \mathbb{N}$ (see **Figure 1**). Since the problem has to be periodic in the "x" direction with the period of the applied field $\frac{2\pi}{\lambda}$, it is enough to analyze only the region where $0 < x < \frac{2\pi}{\lambda}$. The symmetry of the external field $\mathbf{B_0}$, the symmetric boundary conditions on $u(x,z)$, and the boundary conditions on $b(x,z)$ listed above imply that

$$\begin{cases} u\left(\frac{\pi}{2\lambda} - \alpha, z\right) = u\left(\frac{\pi}{2\lambda} + \alpha, z\right) \\ u\left(\frac{3\pi}{2\lambda} - \alpha, z\right) = u\left(\frac{3\pi}{2\lambda} + \alpha, z\right) \text{ and} \\ u\left(\frac{\pi}{\lambda} - \alpha, z\right) = u\left(\frac{\pi}{\lambda} + \alpha, z\right) \end{cases} \begin{cases} b\left(\frac{\pi}{2\lambda} - \alpha, z\right) = -b\left(\frac{\pi}{2\lambda} + \alpha, z\right) \\ b\left(\frac{3\pi}{2\lambda} - \alpha, z\right) = -b\left(\frac{3\pi}{2\lambda} + \alpha, z\right) \\ b\left(\frac{\pi}{\lambda} - \alpha, z\right) = b\left(\frac{\pi}{\lambda} + \alpha, z\right) \end{cases} \tag{13}$$

for any $\alpha \in \mathbb{R}$. One can see now the precise analogy between the flat and spherical cases. In the spherical problem, the meridional angle "ϑ" corresponds to the "x" coordinate in the flat case, with the equatorial plane corresponding to the planes $x = \frac{\pi}{2\lambda}$ and $x = \frac{3\pi}{2\lambda}$. The azimuthal angle "$\varphi$" and the radial coordinate "r" are of course analogous to "y" and "z," respectively.

It is also clear from (13) that the "z" component of the currents $j_z = \frac{\partial b}{\partial x}$ has to be symmetric with respect to the planes $x = \frac{\pi}{2\lambda}, \frac{3\pi}{2\lambda}$ while the "x" component $j_x = -\frac{\partial b}{\partial z}$ remains antisymmetric. This means that j_z must have an external value and j_x must vanish at $x = \frac{\pi}{2\lambda}, \frac{3\pi}{2\lambda}$.

The main flow is defined as the flow outside all boundary and internal layers in the problem. When the upper boundary is insulating or only weakly conducting, the problem is greatly simplified since the magnetic coupling of the fluid with the lower conductor, in the limit of the large Hartmann number, is much stronger than with the upper one. The fluid therefore should lock on to the lower boundary generating large shear in a Hartmann boundary layer adjacent to the upper conductor, where the velocity decreases to zero on a distance in the order of M^{-1}. This allows to deduce that the electrical currents in the system, generated by the flow and circulating through the boundary layers and both boundaries, in particular through the upper poor conductor, should scale as $O(M^{-1})$ everywhere except for the boundary layers where the shear is large. A schematic picture of the current circulation when the outer/upper boundary is poorly conducting or insulating for both geometries is provided in **Figure 3**. It follows, that everywhere the perturbation magnetic field is weak, that is $b(x,z) = O(M^{-1})$. Therefore, from (8), we infer that the main flow for the case of poorly conducting or insulating boundary is determined by

$$\begin{cases} \mathbf{B_0} \cdot \nabla u = 0 + O(M^{-1}) \\ \mathbf{B_0} \cdot \nabla b = 0 + O(M^{-2}) \end{cases} \Rightarrow \begin{cases} u = \mathcal{F}(A) + O(M^{-1}) \\ b = \mathcal{G}(A) + O(M^{-2}) \end{cases} \tag{14}$$

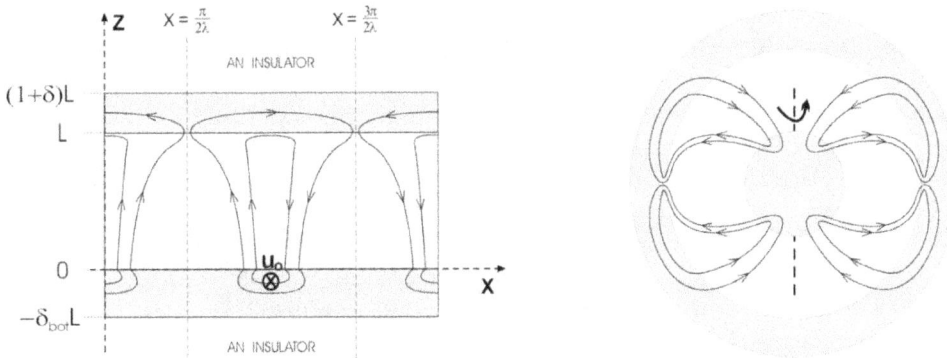

Figure 3. A schematic picture of the current circulation in the planar (left panel) and spherical (right panel) configurations. In the planar case, the direction of the external field oscillates in the x direction and so does the direction of currents, which are strongest in a shear layer along the critical $\mathcal{C}$-line. In the spherical case, strong currents flow from the inner sphere to the outer shell in the shear layer along $\mathcal{C}$ and return in polar regions.

where $\mathcal{F}$ and $\mathcal{G}$ depend on A alone, thus the flow and the induced magnetic field are constant on the field lines. The second equation in (14) means of course, that at the leading order, the Lorentz force vanishes everywhere in the main flow, which in turn implies that the currents are parallel to the external field $\mathbf{B_0}$.

It is clear now that the magnetic field lines which are tangent to the upper boundary, referred to as the $\mathcal{C}$ lines, divide the flow into three regions **I**, **II**, and **III** (see **Figure 1**), and the properties of the solution for each region are somewhat different since in regions **II** and **III**, the external field lines intersect with both boundaries and in region **I**, only with the lower one. Regions **II** and **III** are, therefore, very similar and the only difference between them is the sign of the perturbation magnetic field since it is antisymmetric with respect to the planes $x = \frac{\pi}{2\lambda}, \frac{3\pi}{2\lambda}$. This antisymmetry of b and symmetry of u, together with (14) results in

$$\begin{cases} u \equiv 1 + O\left(M^{-1}\right) \\ b \equiv 0 + O\left(M^{-2}\right) \end{cases} \text{in region I} \qquad (15)$$

thus, the fluid in region **I** flows with the same uniform velocity of the bottom boundary and the perturbation magnetic field vanishes at leading order. Therefore, in region **I**, the currents also must vanish. However, in the area of singularity of the Hartmann layer, namely at $x = \frac{\pi}{2\lambda}, \frac{3\pi}{2\lambda}$ and at $z = 1$, the "z" component of the currents, as it was stated earlier, has an external value (while the "x" component is zero) and interacting with the external magnetic field creates a Lorentz force which has to either accelerate or decelerate the fluid depending on whether the signs of j_z and B_{0x} are the same or opposite.

Since the $\mathcal{C}$ lines that create the singularity of the upper Hartmann layer connect regions of different flow characteristics, a thin area along the lines has to be treated differently and the dissipation must play an important role in this region. Those are shear layers, for which the precise analysis allows to compute the magnitude of super velocity.

The situation is more complicated when the upper boundary is strongly conducting. According to Soward and Dormy [6], the Ferraro's law still holds in region **I** (equatorial region $\mathcal{E}$ for the spherical case) where the fluid locks on to the moving boundary, however, in regions **II** and **III** (equivalently in polar regions $\mathcal{P}$), the Ferraro's law is violated by the influence of ohmic diffusion. This happens, because the induced magnetic field b is no longer small, but of comparable magnitude with the velocity field. Nevertheless, the shear layer along the critical C-line still forms and strong currents enter the upper boundary in the region of tangent contact between the line C and the boundary, thus creating strong Lorentz force, which accelerates the flow. The results of numerical simulations of Mizerski and Bajer [5] are recalled here on **Figure 4** to demonstrate the enhancement of super velocities with the increasing conductivity of the upper boundary.

The obvious conclusion of the above analysis is that the acceleration of the fluid at $x = \pi/2\lambda$ and $z \approx 1$ is due to the curvature of the applied field $\mathbf{B_0}$ generating singularity at this point, and the antisymmetry of external field's "z" component with respect to the plane $x = \pi/2\lambda$ which is responsible for the direction of the currents and therefore also the Lorentz force at $z = 1$. At the singular point, the intensity of the currents entering the boundary layer and the upper boundary increases with the conductivity of this boundary because its interaction with the conducting fluid strengthens. This also implies the increase of the magnitude of the super velocities with ϵ.

These conclusions are also true for the spherical case for which the whole analysis differs only with slightly more complicated boundary conditions and diffusive terms. This complication, however, at the leading order affects mainly the analysis of the shear layer presented in the next section but does not make the main flow analysis more difficult in any way.

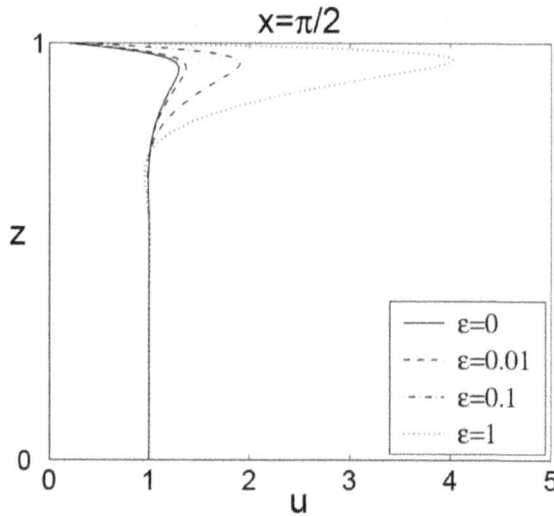

Figure 4. Velocity profiles at $x = \pi/2$ for four different values of the conductivity ratio for the upper boundary $\epsilon = 0$, 0.01, 0.1, and 1. The magnitude of super velocities is the highest near the upper boundary (in the region of singularity of the Hartmann layer) and significantly increases with ϵ. After Mizerski and Bajer [5].

It may also be interesting to make a comment on a similar problem studied numerically by Hollerbach & Skinner [16] of spherical Couette flow with axial magnetic field aligned with the axis of rotation in terms of the singular perturbation method for large Hartmann numbers, infinitesimal rotation and conductivity of the inner sphere. In this case, the Hartmann layers also become singular at the equator where the external filed becomes tangent to the boundaries. This time, however, only the singularity at the inner sphere is important since the field lines tangent to the outer shell leave the fluid and do not couple it to the boundary. Outside a cylinder tangent to the inner sphere and aligned with the axis of rotation the fluid must be at rest, since the velocity field must be constant on the magnetic field lines and the outer stationary sphere has the same conductivity as the fluid, thus the fluid is locked on to it. In such a case, the currents leaving the inner boundary layer at $\vartheta = \frac{\pi}{2}$ and interacting with the external magnetic field create a Lorentz force which decelerates the fluid and produce a counter-rotating jet as found by Hollerbach and Skinner [16].

A simple conclusion which can be stated now is that super- and counter-rotating jets in such MHD systems as considered above are, in general, the outcome of three major features of these systems: the presence of isolated singular points where the external magnetic field is tangent to the boundary, the symmetries of the external magnetic field in respect to planes containing the singular points and perpendicular to the boundaries (namely, antisymmetry of the component perpendicular to the boundary and symmetry of the parallel component) and the symmetric boundary conditions for the velocity field. However, as observed by [10] the singular points can also be created in side the domain (away from the boundaries) by a magnetic field configuration with X-type null points (see field configuration 4 in [10]); also in this case the presence of super-rotation depends on the conductivity of boundaries.

3. The shear layer along the $\mathcal{C}$-line

We will now briefly introduce the reader into the mathematical approach to the analysis of the shear layer structure, which is based on the singular perturbation theory. To take into account of the curvature of the $\mathcal{C}$-line, it is more suitable to use different variables. As mentioned, the symmetries of the system imply that it is enough to limit the analysis to the interval $0 < x < \pi/2\lambda$. Thus, the magnetic field lines can be represented parametrically in the following way

$$x(\tau) = \lambda A(\tau - 1) + \frac{\pi}{2\lambda}$$

$$\exp[\lambda z(\tau)] = \frac{1}{\lambda A} \sin\left[\lambda^2 A(\tau - 1) + \frac{\pi}{2}\right] \tag{16}$$

and the point $x = \frac{\pi}{2\lambda}, z = 1$ where the Hartman layer singularity occurs (referred to as the point $\mathcal{S}$ (see **Figure 5**)) is defined by $A = A_{\mathcal{C}}$ and $\tau = 1$. Introducing a measure of distance along the critical $\mathcal{C}$-line

$$\gamma(\tau) = -\int_{\mathcal{S}} \mathbf{B} \cdot d\mathbf{r} = -\int_{\mathcal{S}} d\Phi = \left. \frac{e^{-\lambda z(\tau')}}{\lambda} \cos[\lambda x(\tau')]\right|_1^\tau = -A \tan[\lambda^2 A(\tau - 1)], \tag{17}$$

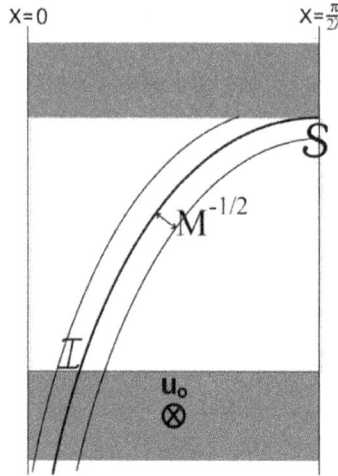

Figure 5. The shear layer along the critical magnetic field line $\mathcal{C}$. The point $\mathcal{S}$ is the point $x = \frac{\pi}{2\lambda}$, $z = 1$ at which the line grazes the upper boundary and the point of intersection of the critical line with the lower boundary is denoted by $\mathcal{I}$. After Mizerski and Bajer [5].

and letting $\Gamma = \gamma(\tau_{\mathcal{I}})$ be the distance between point $\mathcal{S}$ and the point of intersection of the field line $\mathcal{C}$ with the lower boundary (referred to as the point $\mathcal{I}$ (see **Figure 5**)) we introduce a new set of coordinates

$$l = 1 - \frac{\gamma(\tau)}{\Gamma} = 1 - \frac{e^{-\lambda z}}{\lambda \Gamma} \cos(\lambda x) \tag{18}$$

$$n = M^{\frac{1}{2}}\sqrt{\Gamma}(A_C - A) = M^{\frac{1}{2}}\sqrt{\Gamma}\left(A_C - \frac{e^{-\lambda z}}{\lambda} \sin(\lambda x)\right) \tag{19}$$

where l is the coordinate along the basic magnetic field lines which has the property that $l = 1$ at $\mathcal{S}$ and $l = 0$ at $\mathcal{I}$, whereas n is a measure of distance between other field lines and the $\mathcal{C}$-line within the shear layer of thickness $M^{-1/2}$ (cf. [19, 20]). The so-defined coordinate n has the properties, that it is 0 on the $\mathcal{C}$-line, positive in region **II** and negative in region **I**; moreover $A_C = \exp(-\lambda)/\lambda = \sqrt{1 - \lambda^2 \Gamma^2}/\lambda$.

For the simplest case of poorly conducting or insulating upper/outer boundary, with the use of the shear layer coordinates (l, n), the Eq. (8) can now be written at the leading order in the form

$$\frac{\partial V_\pm}{\partial l} \pm \frac{\partial^2 V_\pm}{\partial n^2} = 0 \tag{20}$$

where

$$V_\pm = u \pm Mb. \tag{21}$$

In the spherical geometry, the analogous formulation leads to the coupling of the two equations for $V_\pm$ through nonzero curvature terms on the right-hand sides

$$\frac{\partial V_\pm}{\partial l} \pm \frac{\partial^2 V_\pm}{\partial n^2} = \frac{1}{s_c}\frac{ds_c}{dl} V_\mp. \tag{22}$$

The coupling term, however, may be neglected if the narrow gap limit is assumed. More importantly, however, the two equations, in both—planar and spherical configurations, are coupled through the boundary conditions at $l = 0, 1$, thus at the points $\mathcal{I}$ and $\mathcal{S}$. Eqs. (20) and (22) are diffusion equations (with a source in the spherical case) valid for all $\infty < n < \infty$, with the variable l corresponding to time variable from standard diffusion processes in the case of V_- and $1 - l$ corresponding to time in the case of V_+. The following solving procedure of Eqs. (20) or (22) can be applied. One can utilize the Green's formula for the diffusion equation and first solve for V_+ by the use of the "initial condition" at $l = 1$ ($1 - l = 0$). Then introduce the obtained expression for V_+ into the "initial condition" for V_- at $l = 0$ and utilize the Green's formula again. Finally, matching the two solutions through the condition at $l = 1$ again yields an integral equation for the super velocities. Such procedure leads to an integral equation of Fredholm type, which has been solved numerically in Dormy et al. [4] and Mizerski and Bajer [5] for the two cases $\epsilon = 0$ and $\epsilon = M^{-1}$.

However, when the boundaries are perfectly conducting, the problem becomes more complicated. The same equations as (20) and (22) are obtained for

$$V_\pm = M^{-1/2}u \pm M^{1/2}b. \tag{23}$$

(cf. Eq. (3.20) in [6]), but the problem becomes analytically intractable due to the complications arising from vanishing of the current component parallel to the boundary at $l = 1$. Nevertheless, Soward and Dormy [6] have managed to show that for perfectly conducting boundaries, the strong current leakage from the shear layer into the outer boundary in the vicinity of the critical point causes strong super rotation $\Omega = \mathcal{O}\left(M^{1/2}\right)$.

4. Summary

The plane and spherical magnetohydrodynamic Couette flow with an applied strong external magnetic field creating Hartmann layer singularities on a boundary is a setting where fastly moving jets form, with the magnitude of the flow exceeding that of the moving boundary, which drives the entire flow. These are the so-called super velocities (super rotation in the spherical case). We have concentrated here on the review of the results and analytic approach to the problem of the formation of super velocities in strong, potential fields, with particular emphasis on the enhancement of super velocities by the conductivity of the resting boundary.

As found by Soward and Dormy [6], the conductivity of the resting (upper/outer) boundary ε greatly influences the current leakage from the shear layer to that boundary near the point of its tangent contact with the critical $\mathcal{C}$-line. In the case of weakly conducting boundary $\epsilon M^{3/4} \ll 1$, the current leakage is of the order $\mathcal{O}\left(\epsilon M^{3/4}\right)$ and it increases with ε to become order unity when $\epsilon M^{3/4} \gg 1$. This strong current is perpendicular to the external field in the singular region, and thus a strong Lorentz force $\mathbf{j} \times \mathbf{B}_0$ is created, which accelerates the flow (or decelerates in some cases as shown by [16], depending on the symmetries of the applied field). In the case of interest when the moving boundary is strongly conducting and $M \gg 1$, the resulting super-velocity scales like $\mathcal{O}\left(M^{1/2}\right)$ when $\epsilon \gg 1$ is of the order $\mathcal{O}\left(\epsilon^{2/3}M^{1/2}\right)$ when $1 \gg \epsilon \gg M^{-3/4}$ and becomes order unity when $\epsilon \ll M^{-3/4}$.

Author details

Krzysztof Mizerski

Address all correspondence to: kamiz@igf.edu.pl

Institute of Geophysics, Polish Academy of Sciences, Warsaw, Poland

References

[1] Dormy E, Cardin P, Jault D. MHD flow in a slightly differentially rotating spherical shell, with conducting inner core, in a dipolar magnetic field. Earth and Planetary Science Letters. 1998;**160**:15-30

[2] Hollerbach R. Magnetohydrodynamic flows in spherical shells. In: Egbers C, Pfister G, editors. Physics of Rotating Fluids. Vol. 549. Lecture Notes in Physics. Berlin, Germany: Springer; 2000. pp. 295-316

[3] Hollerbach R. Super- and counter-rotating jets and vortices in strongly magnetic spherical Couette flow. In: Chossat P, Armbruster D, Oprea I, editors. Dynamo and Dynamics, A Mathematical Challenge. Vol. 26. NATO Science Series II. Dordrecht: Kluwer; 2001. pp. 189-197

[4] Dormy E, Jault D, Soward AM. A super-rotating shear layer in magnetohydrodynamic spherical Couette flow. Journal of Fluid Mechanics. 2002;**452**:263-291

[5] Mizerski KA, Bajer K. On the effect of mantle conductivity on super-rotating jets near the liquid core surface. Physics of the Earth and Planetary Interiors. 2007;**160**:245-268

[6] Soward AM, Dormy E. Shear layers in magnetohydrodynamic spherical Couette flow with conducting walls. Journal of Fluid Mechanics. 2010;**645**:145-185

[7] Nataf H-C, Alboussière T, Brito D, Cardin P, Gagnière N, Jault D, Masson J-P, Schmitt D. Experimental study of super-rotation in a magnetostrophic spherical Couette flow, Geophysical and Astrophysical Fluid Dynamics. 2006;**100**:281-298

[8] Brito D, Alboussière T, Cardin P, Gagnière N, Jault D, La Rizza P, Masson J-P, Nataf H-C, Schmitt D. Zonal shear and super-rotation in a magnetized spherical Couette-flow experiment. Physical Review E. 2011;**83**:066310

[9] Wei X, Hollerbach R. Magnetic spherical Couette flow in linear combinations of axial and dipolar fields. Acta Mechanica. 2010;**215**:1-8

[10] Hollerbach R, Hulot D. Shercliff layers in strongly magnetic cylindrical Taylor-Couette flow. Comptes Rendus Mécanique. 2016;**344**:502-509

[11] Song X, Richards PG. Seismological evidence for differential rotation of the Earth's inner core. Nature. 1996;**382**:221-224

[12] Su W, Dziewonski AM, Jeanloz R. Planet within a planet: Rotation of the inner core of earth. Science. 1996;**274**:1883-1887

[13] Alexandrescu M, Gibert D, Le Mouël J-L, Hulot G, Saracco G. An estimate of average lower mantle conductivity by wavelet analysis of geomagnetic jerks. Journal of Geophysical Research. 1999;**104**(B8):17735-17745

[14] Dubrovinsky L, Dubrovinskaia N, Langenhorst F, Dobson D, Rubie D, Geβmann C, Abrikosov IA, Johansson B, Baykov VI, Vitos L, Le Bihan T, Crichton WA, Dmitriev V, Weber H-P. Iron-silica interaction at extreme conditions and the electrically conducting layer at the base of Earth's mantle. Nature. 2003;**422**:58-61

[15] Holme R. Electromagnetic core – mantle coupling – I. Explaining decadal changes in the length of day. Geophysical Journal International. 1998;**132**:167-180

[16] Hollerbach R, Skinner S. Instabilities of magnetically induced shear layers and jets. Proceedings of the Royal Society of London A. 2001;**457**:785-802

[17] Hollerbach R. Magnetohydrodynamic Ekman and Stewartson layers in a rotating spherical shell. Proceedings of the Royal Society of London A. 1994;**444**:333-346

[18] Kleeorin N, Rogachevskii L, Ruzmaikin A, Soward AM, Starchenko S. Axisymetric flow between diferentially rotating spheres in a dipole magnetic field. Journal of Fluid Mechanics. 1997;**344**:213-244

[19] Starchenko SV. Magnetohydrodynamic flow between insulating shells rotating in strong potential field. Physics of Fluids. 1998;**10**:2412-2420

[20] Starchenko SV. Strong potential field influence on slightly differentially rotating spherical shells. Studia Geophysica et Geodaetica. 1998;**42**:314-319

The Heat and Mass Transfer Analysis During Bunch Coating of a Stretching Cylinder by Casson Fluid

Taza Gul and Shakeela Afridi

Additional information is available at the end of the chapter

http://dx.doi.org/10.5772/intechopen.79772

Abstract

The aim of this study is to coat a stretching cylinder with the help of a liquid film spray. The Casson fluid has been chosen for the coating phenomena. The thickness of the liquid film has been used as variable, and the influence of heat and mass transmission under the impact of thermophoresis has been encountered in the flow field. The required pressure term for the spray pattern during variable thickness has mainly been focused. Using the suitable similarity transformations, the basic flow equations for the fluid motion have been converted into high-order nonlinear coupled differential equations. Series solutions of subsequent problem have been obtained using controlling procedure optimal approach. Important physical constraints of skin friction, Nusselt number, and Sherwood number have been calculated numerically and discussed. Other physical parameters involved in the problem, i.e., Reynolds number Re, Casson fluid parameter β_1, Prandtl number Pr, Lewis number Le, Brownian motion parameter N_b, and thermophoresis parameter N_t have been illustrated. The skin friction effect and its physical appearance are also included in this work. The convergence is checked by plotting h-curves. The emerging parameters are discussed by plotting graphs. The recent work is also compared with the published work.

Keywords: thin film spray, Casson nanofluid, stretched cylinder, heat and mass transfer, thermophoresis, HAM

1. Introduction

The relation among pressure and flow is an important phenomena which plays a vital role to understand the circulation of the blood in the human body and its sustainability. The approach of pressure [1] can bring a partial barrier in some areas of the smaller vessels due to the thrilling change of the yield stress. The blood is equally a mixture of two fluids Casson and the other one is Newtonian fluid and studied by Srivastava and Saxena [2]. They focused on the effect of resistance created by the viscosity term and wall shear stresses. Later on, this fluid is studied by many researchers on the stretching surfaces for other industrial and engineering usages [3]. Mahdy [4] have examined the fluid motion over an extending cylinder considering Casson fluid. Hayat et al. [5] have examined the third-order fluid motion over an extending tube within the effect of MHD. Qasim et al. [6] have considered the slip flow of sighted ferrofluid over an extending cylinder. Sheikholeslami [7] has studied the suction idea considering nanofluid over an extended cylinder. Manjunatha et al. [8] have examined the radiation effect in a porous space using dusty fluid and stretching cylinder. Abdulhameed et al. [9] have examined the oscillatory flow phenomena using circular cylinder. Hakeem et al. [10] have studied the flow of Walter's B fluid over an extending sheet. Pandey et al. [11] have studied the Walter's B viscoelastic nanofluid film energetic from below. The interesting and fruitful applications of thin film are the wire and fiber coating, processing of food stuff, extrusion of polymer and metal, drawing of plastic sheets, continuous casting, fluidization of reactor, and chemical processing equipment. On the basis of these applications, researchers did a lot of work on it. Wang [12] was the main researcher who investigated liquid film on an unstable extending surface. Recently Tawade et al. [13] have investigated liquid film flow over an unstable extending surface with thermal radiation, in the existence of continuous magnetic field using numerical method. The liquid film flow considering non-Newtonian fluids proliferates in many life geographies which is used mostly in cylindrical shapes. Several researchers [14–17] investigated power-law fluid with unsteady extending surface using different cases. Megahe [18] and Abolbashari et al. [19] have scrutinized thin film flow of Casson fluid using slip boundary conditions. Recently Qasim et al. [20] have examined the liquid film flow of nanofluid considering Buongiorno's model.

The liquid film spray on a stretching sheet has also an important phenomena to coat the metals and increase their life. The idea of spray on the stretching surface is the study of Wang [21]. Recently Noor et al. [22] considered the thin film spray of nanofluid on a stretching cylinder. They compared their results with the experimental data and found the impact of the physical parameters during flow phenomena. They also discussed the application of their work in detail. Most of the mathematical problems in the field of engineering are composite in their nature, and the exact resolution is very tough or even not conceivable. The solution of these problems is tackled through numerical and analytical methods. Homotopy analysis method is one of the popular techniques for the solution of such complex problems. Liao [22–26] investigates this series solution technique for the solution of nonlinear problems. The other important feature of this method is that its solution contains all the embedded parameters involved in the problem and also the range of the embedded parameters. The high nonlinear problems have been solved by Abbasbandy [27] due to the fast convergence of this method. Alshomrani and Gul [28], Gul [29] have studied the solution of nonlinear differential equations through HAM arises in the field of engineering and industry.

2. Formulation

Consider the thin film flow of Casson nanofluid elegantly through a circular cylinder of radius "a." The cylinder is supposed to be stretched along with radial direction with velocity U_w, and temperature at the surface of cylinder is taken T_w. The uniform ambient temperature is considered T_b such that $T_w - T_b > 0$ for assisting flow and $T_w - T_b < 0$ for opposing flow.

The governing equations of continuity, heat transfer, and mass transfer are

$$\frac{\partial u}{\partial r} + \frac{u}{r} + \frac{\partial w}{\partial z} = 0, \tag{1}$$

$$u\frac{\partial w}{\partial r} + w\frac{\partial w}{\partial z} = v\left(1 + \frac{1}{\beta_1}\right)\left(\frac{\partial^2 w}{\partial r^2} + \frac{1}{r}\frac{\partial w}{\partial r}\right) + g\beta^*(T - T_b)(1 - C_b) + \frac{1}{\rho}(\rho^* - \rho)(C - C_b), \tag{2}$$

$$u\frac{\partial u}{\partial r} + w\frac{\partial u}{\partial z} = -\frac{1}{\rho}\frac{\partial P}{\partial r} + v\left(1 + \frac{1}{\beta_1}\right)\left(\frac{\partial^2 u}{\partial r^2} + \frac{1}{r}\frac{\partial u}{\partial r} - \frac{u}{r^2}\right), \tag{3}$$

$$u\frac{\partial T}{\partial r} + w\frac{\partial T}{\partial z} = \alpha\left(\frac{\partial^2 T}{\partial r^2} + \frac{1}{r}\frac{\partial T}{\partial r}\right) + \frac{\rho^* c_p^*}{\rho c_p}\left(D_B\frac{\partial T}{\partial r}\frac{\partial \phi}{\partial r} + \frac{D_T}{T_b}\left(\frac{\partial T}{\partial r}\right)^2\right), \tag{4}$$

$$u\frac{\partial C}{\partial r} + w\frac{\partial C}{\partial z} = D_B\left(\frac{\partial^2 C}{\partial r^2} + \frac{1}{r}\frac{\partial C}{\partial r}\right) + \frac{D_T}{T_b}\left(\frac{\partial^2 T}{\partial r^2} + \frac{1}{r}\frac{\partial T}{\partial r}\right), \tag{5}$$

where $u(r,z)$ and $w(r,z)$ are velocity components; ρ is density; v is kinematic viscosity; β_1 is the constant characteristic to Casson fluid; β^* is the coefficient of thermal expansion; g is the gravitational acceleration along z-axis; T, T_b, C and C_b determine the temperature, ambient temperature, concentration, and ambient concentration, respectively; and α, D_T, D_B stands for the thermal diffusivity, thermophoresis diffusion coefficient, and Brownian diffusion coefficient.

The suitable boundary conditions are

$$u = U_w, w = W_w, T = T_w, C = C_w \quad at \quad r = a, \tag{6}$$

$$\mu\frac{\partial w}{\partial r} = \frac{\partial T}{\partial r} = \frac{\partial C}{\partial r} = 0, u = w\frac{d\delta}{dz} \quad at \quad r = b. \tag{7}$$

Here $U_w = -ca$ represents the suction and injection velocity, and $W_w = 2cz$ is the stretching velocity such that c represents the stretching parameter and δ is the thickness of fluid film.

The similarity transformations are used to alter the basic Eqs. (1)–(7) used in [22] as

$$u = -ca\frac{f(\eta)}{\sqrt{\eta}}, w = 2cz\frac{df}{d\eta}, T(z) = T_b - T_{ref}\left(\frac{cz^2}{v_{nf}}\right)\theta(\eta), C(z) = C_b - C_{ref}\left(\frac{cz^2}{v_{nf}}\right)\phi(\eta), \tag{8}$$

where

$$\eta = \left(\frac{r}{a}\right)^2.$$

In the case of the outer radius b of the flow, $\eta = \left(\frac{b}{a}\right)^2$.

Using these transformations in Eqs. (1), (2), (4)–(7), we obtained a set of dimensionless equations which is

$$\left(1+\frac{1}{\beta_1}\right)\left(\eta\frac{\partial^3 f}{\partial\eta^3}+\frac{\partial^2 f}{\partial\eta^2}\right)+\text{Re}\left(f\frac{\partial^2 f}{\partial\eta^2}-\left(\frac{\partial f}{\partial\eta}\right)^2+\lambda(\theta+Nr\phi)\right)=0, \tag{9}$$

$$\eta\frac{\partial^2\theta}{\partial\eta^2}+\frac{\partial\theta}{\partial\eta}+\text{Pr}.\text{Re}\left(f\frac{\partial\theta}{\partial\eta}-2\frac{\partial f}{\partial\eta}\theta\right)+\eta\frac{\partial\theta}{\partial\eta}\left(N_t\frac{\partial\theta}{\partial\eta}+N_b\frac{\partial\phi}{\partial\eta}\right)=0, \tag{10}$$

$$\eta\frac{\partial^2\phi}{\partial\eta^2}+\frac{\partial\phi}{\partial\eta}+Le.\text{Re}\left(f\frac{\partial\phi}{\partial\eta}-2\frac{\partial f}{\partial\eta}\phi\right)+\frac{N_t}{N_b}\left(\eta\frac{\partial^2\theta}{\partial\eta^2}+\frac{\partial\theta}{\partial\eta}\right)=0, \tag{11}$$

where

$$\text{Re}=\frac{ca^2}{2v_{nf}},\ \lambda=\frac{g\beta^* a(T_w-T_\infty)(1-C_\infty)}{W_w^2},\ Nr=\frac{(\rho-\rho^*)(C_w-C_\infty)}{\rho\beta^*(T_w-T_\infty)(1-C_\infty)},$$

$$\text{Pr}=\frac{\mu c_p}{k},\ N_t=\frac{\rho^* c_p^* D_T\Delta T}{\rho c_p\alpha T_b},\ Le=\frac{v}{D_B},\ N_b=\frac{\rho^* c_p^* D_B\Delta C}{\rho c_p\alpha}. \tag{12}$$

In Eq. (12) Re stands for the Reynolds number, λ is the buoyancy parameter or in other word it is the natural convection parameter, Nr stands for the buoyancy ratio, Pr represents the Prandtl number, N_t is used to represent thermophoresis parameter, Le is Lewis number, and N_b is Brownian motion parameter.

Physical conditions for momentum, thermal, and concentration fields are transformed as

$$f(1)=\frac{\partial f(1)}{\partial\eta}=\theta(1)=\phi(1)=1, \tag{13}$$

$$\frac{\partial^2 f(\beta)}{\partial\eta^2}=f(\beta)=\frac{\partial\theta(\beta)}{\partial\eta}=\frac{\partial\phi(\beta)}{\partial\eta}=0, \tag{14}$$

where β is the thickness of liquid film sprayed on the outer surface of the cylinder.

Integrating Eq. (3) for pressure term

$$\frac{p-p_b}{\mu c}=\frac{\text{Re}}{\eta}f^2-2\left(1+\frac{1}{\beta_1}\right)\frac{\partial f}{\partial\eta}. \tag{15}$$

At the outer surface, the shear stress of the liquid film is zero, i.e.,

$$\frac{\partial^2 f(\beta)}{\partial\eta^2}=0. \tag{16}$$

The shear stress on the cylinder is

$$\tau=\frac{\rho v 4cz}{a}f''(1)=\frac{4c\mu z}{a}f''(1) \tag{17}$$

The deposition velocity V is written as

$$-V = -ca\frac{f'(\beta)}{\sqrt{\beta}}. \tag{18}$$

Mass flux m_1 is in association with the deposition per axial length which is

$$m_1 = V2\pi b \tag{19}$$

The normalized mass flux m_2 is

$$m_2 = \frac{m_1}{2\pi a^2 c} = \frac{m_1}{4\pi v_{nf} \mathrm{Re}} = f(\beta) \tag{20}$$

The flow, temperature, and concentration rates are

$$S_f = \frac{2v_{nf}}{W_w}\left(\frac{\partial w}{\partial r}\right)_{r=a}, Nu = -\frac{ak}{(T_w - T_b)}\left(\frac{\partial T}{\partial r}\right)_{r=a}, Sh = -\frac{a}{2(C_w - C_b)}\left(\frac{\partial C}{\partial r}\right)_{r=a}. \tag{21}$$

The nondimensional forms for the abovementioned physical properties are

$$\frac{z\mathrm{Re}}{a}S_f = \frac{\partial^2 f(1)}{\partial \eta^2}, Nu = -2k\frac{\partial\theta(1)}{\partial\eta}, Sh = -\frac{\partial\phi(1)}{\partial\eta}. \tag{22}$$

3. Solution by homotopy analysis method

Initially guessed values for f, θ, and ϕ at $\eta = 1$ are

$$f_0(\eta) = \frac{\beta}{2(\beta - 1)^3}\left[\eta^3 - 3\beta\eta^2 - (3 - 6\beta)\eta + (2 - 3\beta)\right] + \eta,$$
$$\theta_0(\eta) = \frac{-\eta^2}{2} + \beta(\eta - 1) + \frac{3}{2}, \phi_0(\eta) = \frac{-\eta^2}{2} + \beta(\eta - 1) + \frac{3}{2}. \tag{23}$$

The linear operators for the given functions L_f, L_θ and L_ϕ are selected as

$$L_f = \frac{\partial^4 f}{\partial \eta^4}, \quad L_\theta = \frac{\partial^2\theta}{\partial \eta^2} \quad \text{and} \quad L_\phi = \frac{\partial^2\phi}{\partial \eta^2}, \tag{24}$$

which satisfies the following general solution:

$$L_f\left(A_1 + A_2\eta + A_3\eta^2 + A_4\eta^3\right) = 0, \quad L_\theta(A_5 + A_6\eta) = 0 \quad \text{and} \quad L_\phi(A_7 + A_8\eta) = 0, \tag{25}$$

where $A_i(i = 1 - 8)$ are constants of general solution.

The corresponding nonlinear operators N_f, N_θ, and N_ϕ are defined as

$$N_f[f(\xi;p),\theta(\xi;p)] = \left(1+\frac{1}{\beta_1}\right)\left(\eta\frac{\partial^3 f(\eta;p)}{\partial\eta^3}+\frac{\partial^2 f(\eta;p)}{\partial\eta^2}\right)$$

$$+\text{Re}\left(f(\eta;p)\frac{\partial^2 f(\eta;p)}{\partial\eta^2}-\left(\frac{\partial f(\eta;p)}{\partial\eta}\right)^2\right)+\left(\lambda\theta(\eta;p)+Nr\phi(\eta;p)\right)=0, \tag{26}$$

$$N_\theta[f(\xi;p),\theta(\xi;p)] = \eta\frac{\partial^2\theta(\eta;p)}{\partial\eta^2}+\frac{\partial\theta(\eta;p)}{\partial\eta}+\text{Pr}.\text{Re}\left(f(\eta;p)\frac{\partial\theta(\eta;p)}{\partial\eta}-2\frac{\partial f(\eta;p)}{\partial\eta}\theta(\eta;p)\right)$$

$$+\eta\frac{\partial\theta(\eta;p)}{\partial\eta}\left(N_t\frac{\partial\theta(\eta;p)}{\partial\eta}+N_b\frac{\partial\phi(\eta;p)}{\partial\eta}\right)=0, \tag{27}$$

$$N_\phi[f(\eta;p),\phi(\eta;p)] = \eta\frac{\partial^2\phi(\eta;p)}{\partial\eta^2}+\frac{\partial\phi(\eta;p)}{\partial\eta}+Le.\text{Re}\left(f(\eta;p)\frac{\partial\phi(\eta;p)}{\partial\eta}-2\frac{\partial f(\eta;p)}{\partial\eta}\phi(\eta;p)\right)$$

$$+\frac{N_t}{N_b}\left(\eta\frac{\partial^2\theta(\eta;p)}{\partial\eta^2}+\frac{\partial\theta(\eta;p)}{\partial\eta}\right)=0, \tag{28}$$

where $p\in[0,1]$ is embedded parameter.

4. Zeroth-order deformation problem

The equations of zeroth-order deformation problem are obtained as

$$(1-p)L_f[f(\eta;p)-f_0(\eta)]=ph_fN_f[f(\eta;p)], \tag{29}$$

$$(1-p)L_\theta[\theta(\eta;p)-\theta_0(\eta)]=ph_\theta N_\theta[f(\eta;p),\theta(\eta;p)], \tag{30}$$

$$(1-p)L_\phi[\phi(\eta;p)-\phi_0(\eta)]=ph_\phi N_\phi[f(\eta;p),\phi(\eta;p)]. \tag{31}$$

Here h_f, h_θ and h_ϕ are auxiliary nonzero parameters. The corresponding boundary conditions are written as

$$f(\eta;p)_{\eta=1}=1, \left.\frac{\partial f(\eta;p)}{\partial\eta}\right|_{\eta=1}=1, \theta(\eta;p)|_{\eta=1}=1, \phi(\eta;p)|_{\eta=1}=1, \tag{32}$$

$$\left.\frac{\partial^2 f(\eta;p)}{\partial\eta}\right|_{\eta=\beta}=0, \left.\frac{\partial\theta(\eta;p)}{\partial\eta}\right|_{\eta=\beta}=0, \left.\frac{\partial\phi(\eta;p)}{\partial\eta}\right|_{\eta=\beta}=0. \tag{33}$$

Since

$$p=0\Rightarrow f(\eta;0)=f_0(\eta)=\eta, \theta(\eta;0)=\theta_0(\eta)=1, \phi(\eta;0)=\phi_0(\eta)=1, \tag{34}$$

$$p = 1 \Rightarrow f(\eta; 1) = f(\eta),\, \theta(\eta; 1) = \theta(\eta),\, \phi(\eta; 1) = \phi(\eta). \tag{35}$$

Using the Taylor's expansions of $f(\eta; p)$, $\theta(\eta; p)$ and $\phi(\eta; p)$ about $p = 0$ in Eqs. (28)–(31), we obtained

$$f(\eta; p) = f_0(\eta) + \sum_{w=1}^{\infty} f_w(\eta) p^w, \tag{36}$$

$$\theta(\eta; p) = \theta_0(\eta) + \sum_{w=1}^{\infty} \theta_w(\eta) p^w, \tag{37}$$

$$\phi(\eta; p) = \phi_0(\eta) + \sum_{w=1}^{\infty} \phi_w(\eta) p^w, \tag{38}$$

where

$$f_w(\eta) = \frac{1}{w!} \frac{\partial^w f (\eta; p)}{\partial \eta^w}\bigg|_{p=0},\, \theta_w(\eta) = \frac{1}{w!} \frac{\partial^w \theta (\eta; p)}{\partial \eta^w}\bigg|_{p=0},\, \phi_w(\eta) = \frac{1}{w!} \frac{\partial^w \phi (\eta; p)}{\partial \eta^w}\bigg|_{p=0}. \tag{39}$$

The convergence of series depends on h_f, h_θ, and h_φ. So let us suppose that series converges at $p = 1$ for some values of h_f, h_θ, and h_φ, then Eqs. (35)–(37) become

$$f(\eta) = f_0(\eta) + \sum_{w=1}^{\infty} f_w(\eta), \tag{40}$$

$$\theta(\eta) = \theta_0(\eta) + \sum_{w=1}^{\infty} \theta_w(\eta), \tag{41}$$

$$\phi(\eta) = \phi_0(\eta) + \sum_{w=1}^{\infty} \phi_w(\eta). \tag{42}$$

5. wth order deformation problem

By taking w times derivatives of Eqs. (28)–(32) and then dividing by $w!$ as well as substituting $p = 0$, we obtained the following equations:

$$L_f\big[f_w(\eta) - \chi_w f_{w-1}(\eta)\big] = \hbar_f\, \mathfrak{R}_w^f(\eta), \tag{43}$$

$$L_\theta\big[\theta_w(\eta) - \chi_w \theta_{w-1}(\eta)\big] = \hbar_\theta\, \mathfrak{R}_w^f(\eta), \tag{44}$$

$$L_\phi\big[\phi_w(\eta) - \chi_w \phi_{w-1}(\eta)\big] = \hbar_\phi\, \mathfrak{R}_w^f(\eta), \tag{45}$$

where

$$\chi_w = \begin{cases} 0, & \text{if } p \le 1 \\ 1, & \text{if } p > 1 \end{cases}.$$

$$\mathfrak{R}_w^f(\eta) = \left(1 + \frac{1}{\beta_1}\right)(\eta f_{w-1}''' + f_{w-1}'') + \text{Re}\sum_{j=0}^{w-1}\left(f_{w-1-j}f_j'' - f_{w-1-j}'f_j'\right) + (Gr_t\theta_{w-1} + Gr_c\phi_{w-1}),$$

(46)

$$\mathfrak{R}_w^\theta(\eta) = \eta\theta_{w-1}'' + \theta_{w-1}' + \text{Pr.Re}\sum_{j=0}^{w-1}\left(f_{w-1-j}\theta_j' - 2f_{w-1-j}'\theta_j\right) + \eta\theta_{w-1}'\sum_{j=0}^{w-1}\left(N_t\theta_{w-1}' + N_b\phi_{w-1}'\right),$$ (47)

$$\mathfrak{R}_w^\phi(\eta) = \eta\phi_{w-1}'' + \phi_{w-1}' + Le.\text{Re}\sum_{j=0}^{w-1}\left(f_{w-1-j}\phi_j' - 2f_{w-1-j}'\phi_j\right) + \frac{N_t}{N_b}\left(\eta\theta_{w-1}'' + \theta_{w-1}'\right).$$ (48)

The related boundary conditions are

$$f_w(1) = f_w'(1) = \theta_w(1) = \phi_w(1) = 1,$$

$$f_w''(\beta) = \theta_w'(\beta) = \phi_w'(\beta) = 0.$$

(49)

The general solution of Eqs. (42)–(44) is given by

$$f_w(\eta) = e_1 + e_2\eta + e_3\eta^2 + f_w^*(\eta),$$

$$\theta_w(\eta) = e_4 + e_5\eta + \theta_w^*(\eta),$$ (50)

$$\phi_w(\eta) = e_6 + e_7\eta + \phi_w^*(\eta).$$

Here $f_w^*(\xi)$, $g_w^*(\xi)$ and $\theta_w^*(\xi)$ represent the particular solutions, and the constant $A_i (i = 1 - 8)$ are determined from boundary conditions (49).

6. Discussion about graphical results

The purpose of this study is to enhance the heat and mass diffusion by choosing a thin-layer spray of the Casson nanofluid over a stretching cylinder. The physical configuration of the problem is shown in **Figure 1**. The solution of the problem has been obtained using the homotopy approach, and the main features for the convergence (h-curves) of homotopy analysis method (HAM) have been shown in **Figures 2** and **3**. These figures demonstrate the h-curves for velocity, temperature, and concentration fields, respectively. The impact of buoyancy parameter λ and buoyancy ratio Nr on velocity field is prescribed in **Figure 4**. Velocity grows with the rising values of λ because the natural convection parameter λ and momentum boundary layer are in direct relation. The similar effect for the rising values of N_r can be seen in **Figure 4**. The effect of thickness parameter β and Casson fluid parameter β_1 versus velocity field is shown in **Figure 5**. Increasing values of β generate friction force and decline the velocity field because the thicker flow creates hurdles in fluid motion, while the thin layer is comparatively fast flowing.

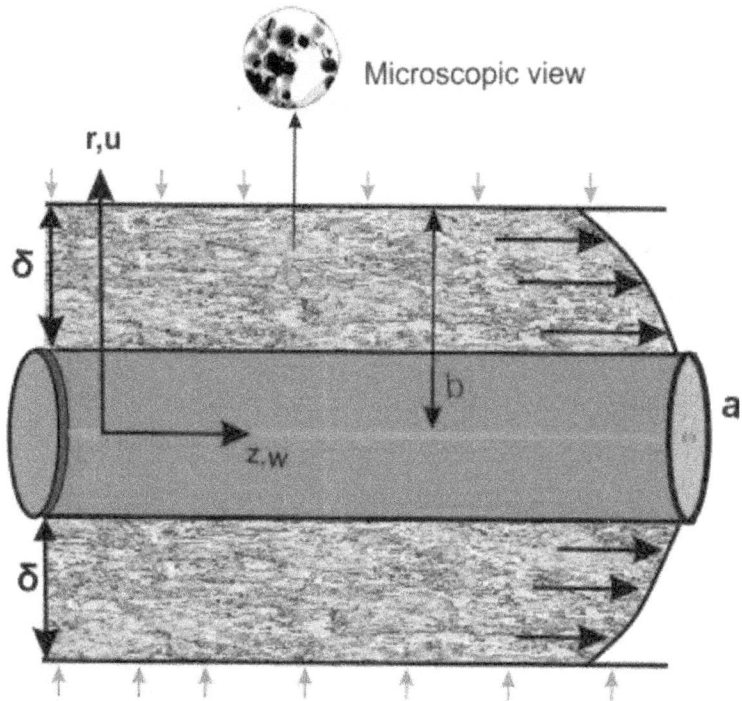

Figure 1. Physical geometry of the problem.

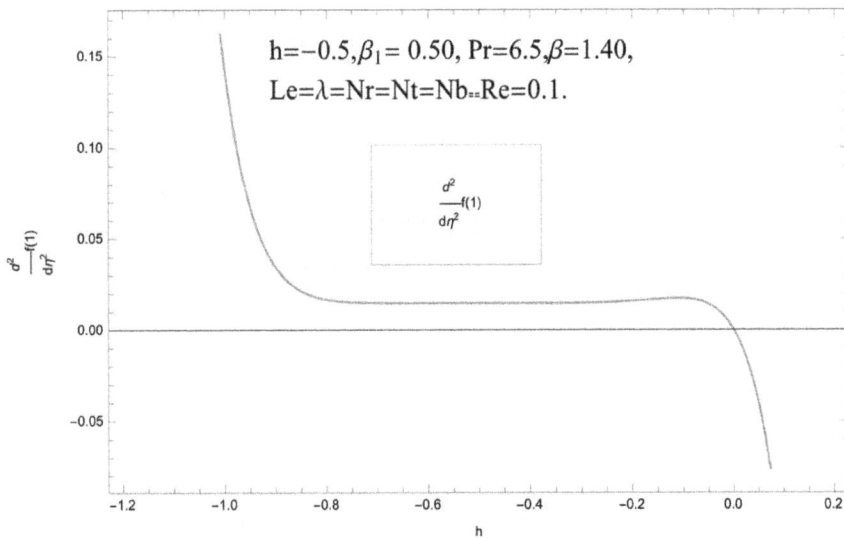

Figure 2. h-curve for velocity profile.

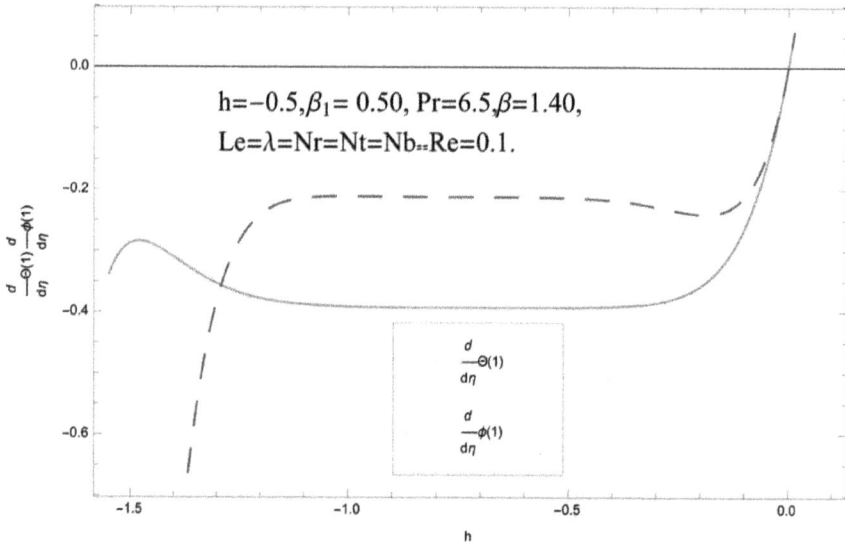

Figure 3. Combined h-curve for temperature and concentration fields.

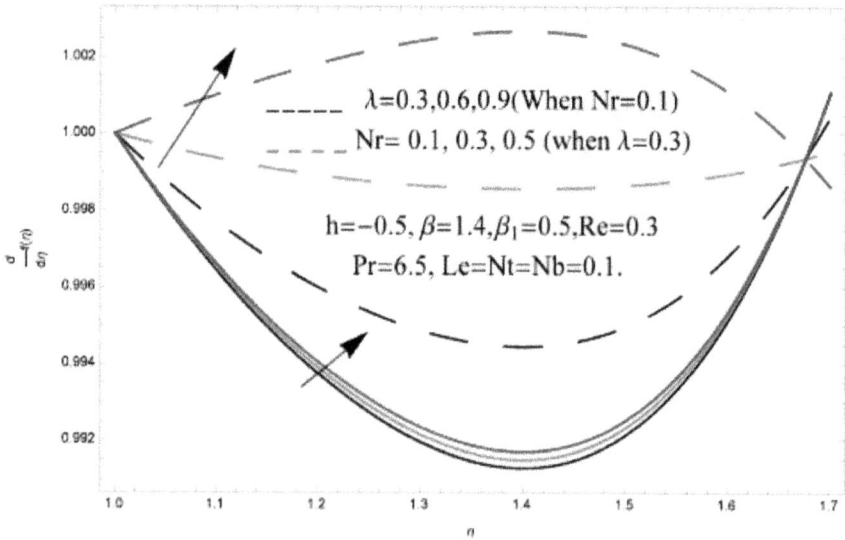

Figure 4. Variation of velocity with Nr and λ.

Therefore, larger amount of β declines the flow motion. The similar effect of the larger amount of the Casson fluid parameter β_1 is shown in **Figure 5**. The rising values of the parameter β_1 imply a decline in the yield stress of the Casson fluid. In **Figure 6**, the behavior of the thermophoretic parameter N_t and Reynolds number Re is observed over the field of temperature.

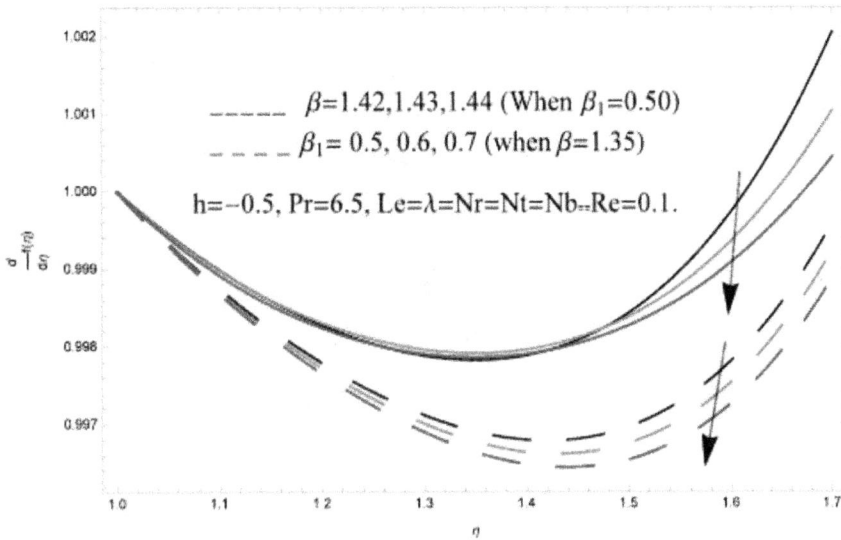

Figure 5. Variation of velocity with β and β_1.

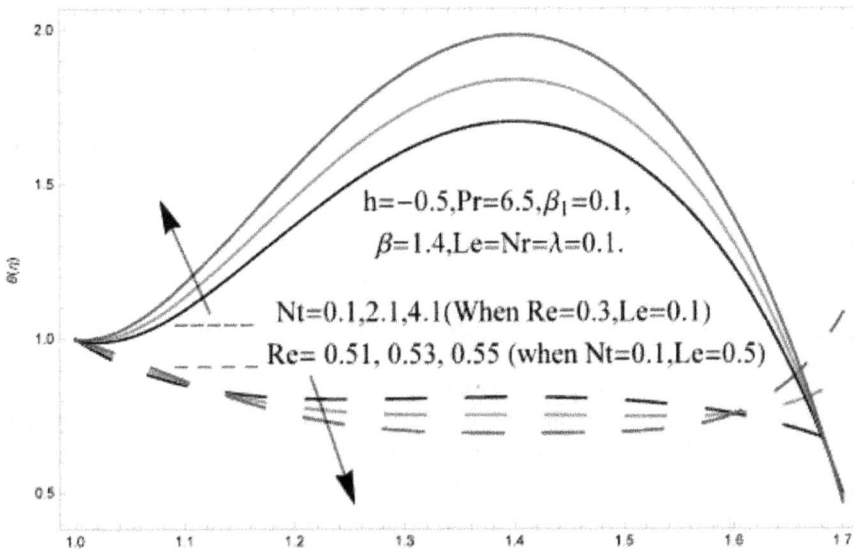

Figure 6. Variation of Nt and Re.

The larger amount of thermophoresis parameter N_t depreciates temperature profile because rising values of N_t enhance the concentration profile due to its direct relation and its product in the model equation increases the cooling effect to reduce the temperature field. The larger quantity of Re reduces temperature field. Rising values of Reynolds number Re enhance the

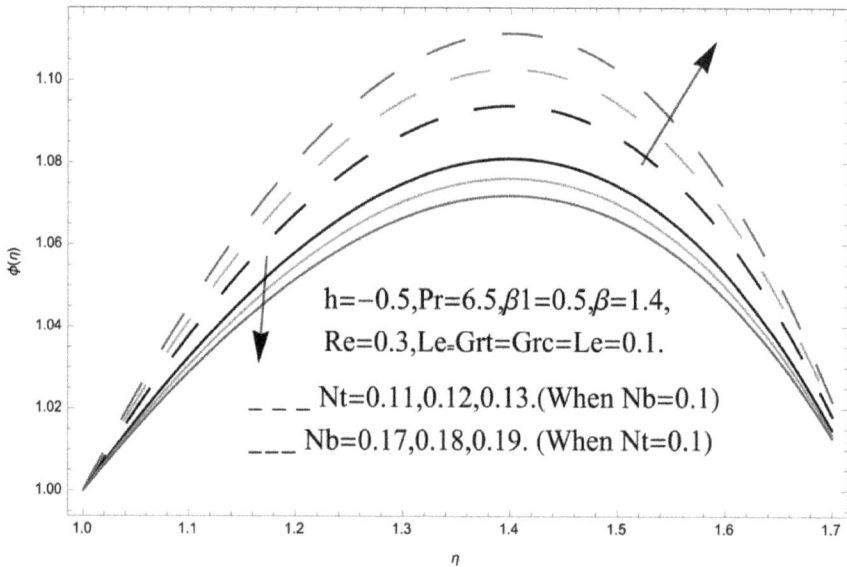

Figure 7. Variation of Nt and Nb.

inertial forces. The powerful inertial forces kept the fluid particles tightly closed, and more heat energy is required to break down the bonds among these atoms. In other words the inertial forces raise the boiling point of the fluid, and more heat energy is required to enhance the temperature. **Figure 7** shows the influences of thermophoretic parameter N_t and Brownian motion parameter N_b in concentration field. The larger amount of N_b displays a falling performance against concentration field. The parameter N_b is owing to the thinning of boundary layer because the random flow of liquid particles makes the decline in the concentration. The rising values of thermophoresis parameter N_t enhance the concentration field. The reason behind this is that N_t is in direct relation with concentration pitch, while the N_b is in inverse relation to the concentration field.

Figure 8 represents the behavior of the concentration field with respect to Reynolds number Re and Lewis number Le. The larger amount of Re improves the concentration field. The reason is that larger values of Re generate the enhancement in the inertial forces to rise concentration field. The concentration boundary layer is falling with the rising value of Lewis number Le.

Figure 9 shows the relationship between pressure distribution over the stretching surface versus Reynolds number Re and Casson fluid parameter β_1. The larger amount of β_1 increases the viscous forces, and more pressure are required at the surface. Thus the larger amount of β_1 decreases the pressure distribution. The larger amount of the Reynolds number Re decreases the pressure distribution. The strong inertial effects packed the fluid particle tightly, and as a result the pressure distribution decreases.

Table 1 shows the numerical values of the skin friction coefficient, local Nusselt number, and Sherwood number of various physical parameters. The skin friction coefficient rises with the

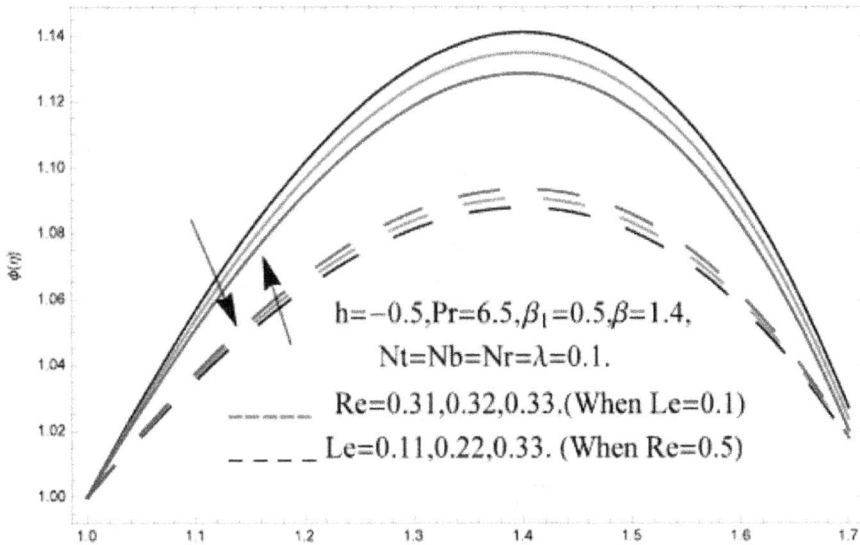

Figure 8. Variation of Re and *Le*.

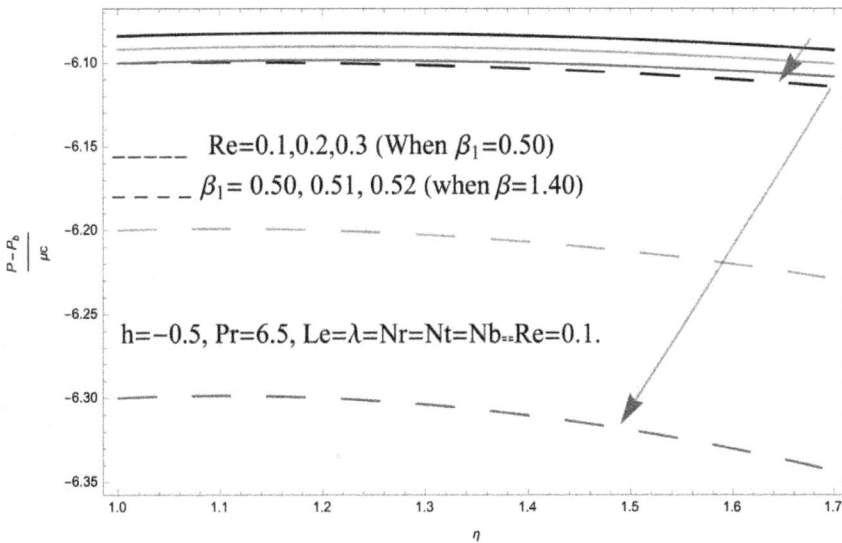

Figure 9. Variation of pressure term.

growth of thickness parameter β. The thick boundary layer increases friction force and improves the cooling effect. Therefore, the Nussselt and Sherwood numbers are increased. The Reynolds number Re decreases the fluid flow due to inertial forces. Due to this reason, the larger quantity of Re enhances the $f''(1), \Theta'(1)$ and $\phi'(1)$. Similar effect for rising values of the

β	Re	β_1	$f''(1)$	$\theta'(1)$	$\phi'(1)$
1.5	0.8	1.2	0.0149342	0.785646	0.212503
1.6			0.0150964	0.916634	0.260143
1.7			0.01673959	1.03338	0.309284
	0.8		0.0149342	0.785646	0.212503
	0.9		0.0151422	0.881567	0.240775
	1.0		0.0149653	0.976928	0.269410
		1.2	0.0149342	0.785646	0.212503
		1.3	0.0154906	0.785691	0.212518
		1.4	0.0160015	0.785733	0.212531

Table 1. Numerical values for skin friction coefficient, local Nusselt number, and Sherwood number for various physical parameters when $h = -0.5, Pr = 0.5, \beta_1 = 1.2, \beta = 1.5, Nt = 0.5, Nb = 1, Nr = 0.6, Re = 0.8, Le = 0.5$.

Re	[30]	[31]	[4]	Present
0.5	0.88220	0.8827	0.88691	0.886942
1.0	1.17776	1.1781	1.17953	1.72926
2.0	1.59390	1.5941	1.59434	3.26714
5.0	2.41745	2.4175	2.4175	6.75053
10.0	3.34445	3.3445	3.34447	10.1078

Table 2. Values of $f''(1)$ for various Reynolds numbers when $h = -0.5, Pr = 0.5, \beta_1 = 1.2, \beta = 1.1, Nt = 0.5, Nb = 1, Nr = 0.6, Le = 0.5$.

Pr	[30]	[31]	[4]	Present
0.7	1.568	1.5683	1.56878	1.56846
2.0	3.035	3.0360	3.03596	3.68121
7.0	6.160	6.1592	6.15813	7.24452
10.0	10.77	7.4668	7.46477	10.2634

Table 3. Values of $-\theta'(1)$ for various Prandtl numbers when $h = -0.5, \beta_1 = 1.2, \beta = 1.1, Nt = 0.5, Nb = 1, Nr = 0.6, Re = 0.8, Le = 0.5$.

Casson parameter β_1 has been shown in **Table 1**. The reason is that the viscous forces become dominant with the larger amount of β_1 to enhance the $f''(1), \Theta'(1)$ and $\phi'(1)$. The comparison of present work and published work has been shown in **Tables 2** and **3**, and closed agreement for $f''(1), \Theta'(1)$ and $\phi'(1)$ has been achieved.

7. Conclusion

The heat and mass transfer effect of a thin film over the extended surface of a cylinder has been explored in the recent research. The spray phenomenon has been studied in the form of velocity, temperature, concentration, and pressure distribution profiles, respectively. The similarity transformation has been used to alter the governing equations into the set of nonlinear differential equations. The solution of the problems has been obtained through the homotopy analysis method (HAM). The impact of the embedded parameters has been examined and discussed. The outcomes of the recent study have been pointed out as:

- The inertial forces become stronger with the larger amount of Reynolds number Re, and as a result the velocity of the fluid flow reduces, while the upsurge values of Re enhance the $f''(1)$. Similarly, the Nusselt number and the Sherwood number are also increasing.

- The larger amount of the thickness parameter β of the thin film produces hurdles in the spray phenomenon, and as a result the velocity of the fluid decreases. On the other hand, the skin friction, Nusselt number, and the Sherwood number grow with the larger values of β. In fact, the cooling effect increases with the rising values of β to enhance the friction force.

- The greater amount of the Brownian motion parameter N_b declines the concentration field. The reason is that the rising values of N_b improve the thinning of the fluid layer and as a result the concentration profile reduces.

- The temperature field increases with the rising value of the thermophoresis parameter N_t, while the concentration field falls to reduce with the larger amount of N_t, because the thermophoresis parameter is in inverse relation to the concentration profile.

- The comparison of the present study with the published work authenticates the obtained result.

Author details

Taza Gul[1,2]* and Shakeela Afridi[1]

*Address all correspondence to: tazagulsafi@yahoo.com

1 Higher Education Department, Khyber Pakhtunkha, Pakistan

2 Department of Mathematics, City University of Science and Information Technology, Pakistan

References

[1] Joly L. In: Copley, editor. Hemorheology. New York: Pergamon Press; 1967. pp. 1-41

[2] Srivastava VP, Saxena M. Two-layered model of casson fluid flow through stenotic blood vessels: Application to the cardiovascular system. Journal of Biomechanics. 1994;11:921-928

[3] Pramanik S. Casson fluid flow and heat transfer past an exponentially porous stretching surface in presence of thermal radiation. Ain Shams Engineering Journal. 2014;5:205-212

[4] Mahdy A. Heat transfer and flow of a casson fluid due to a stretching cylinder with the Soret and Dufour effects. Journal of Engineering Physics and Thermophysics. 2015;**88**:928-936

[5] Hayat T, Shafiq A, Alsaedi A. MHD axisymmetric flow of third grade fluid by a stretching cylinder. Alexandria Engineering Journal. 2015;**54**:205-212

[6] Qasim M, Khan ZH, Khan WA, Shah IA. MHD boundary layer slip flow and heat transfer of ferrofluid along a stretching cylinder with prescribed heat flux. PLoS One. 2014;**9**:1-6

[7] Sheikholeslami M. Effect of uniform suction on nanofluid flow and heat transfer over a cylinder. Journal of the Brazilian Society of Mechanical Sciences and Engineering. 2015;**37**: 1623-1633

[8] Manjunatha PT, Gireesha BJ, Prasannakumara BC. Effect of radiation on flow and heat transfer of MHD dusty fluid over a stretching cylinder embedded in a porous medium in presence of heat Source. International Journal of Applied Computational Science and Mathematics. 2017;**3**:293-310

[9] Abdulhameed M, Vieru D, Sharidan S. Comparison of different pressure waveforms for heat transfer performance of oscillating flow in a circular cylinder. Engineering Science and Technology, an International Journal. 2016;**19**:1040-1049

[10] Hakeem AKA, Ganesh NV, Ganga B. Effect of heat radiation in a Walter's liquid B fluid over a stretching sheet with non-uniform heat source/sink and elastic deformation. Journal of King Saud University—Engineering Sciences. 2014;**26**:168-175

[11] Pandey SD, Nema VK, Tiwari S. Charateristic of Walter's B visco-elastic nanofluid layer heated from below. International Journal of Energy Engineering. 2016;**6**:7-13

[12] Wang CY. Liquid film on an unsteady stretching surface. Quarterly of Applied Mathematics. 1990;**48**:601-610

[13] Tawade L, Abel M, GMetri P, Koti A. Thin film flow and heat transfer over an unsteady stretching sheet with thermal radiation, internal heating in presence of external magnetic field. International Journal of Applied Mathematics and Mechanics. 2016;**3**:29-40

[14] Andersson HI, Aarseth JB, Braud N, Dandapat BS. Flow of a power-law fluid film on an unsteady stretching surface. Journal of Non-Newtonian Fluid Mechanics. 1996;**62**:1-8

[15] Chen CH. Heat transfer in a power-law liquid film over an unsteady stretching sheet. Heat and Mass Transfer. 2003;**39**:791-796

[16] Wang C, Pop L. Analysis of the flow of a power-law liquid film on an unsteady stretching surface by means of homotopy analysis method. Journal of Non-Newtonian Fluid Mechanics. 2006;**138**:161-172

[17] Chen CH. Effect of viscous dissipation on heat transfer in a non-Newtonian liquid film over an unsteady stretching sheet. Journal of Non-Newtonian Fluid Mechanics. 2006;**135**: 128-135

[18] Megahe AM. Effect of slip velocity on Casson thin film flow and heat transfer due to unsteady stretching sheet in presence of variable heat flux and viscous dissipation. Applied Mathematics and Mechanics. 2015;**36**:1273-1284

[19] Abolbashari HM, Freidoonimehr N, Rashidi MM. Analytical modeling of entropy generation for Casson nano-fluid flow induced by a stretching surface. Advanced Powder Technology. 2015;**26**:542-552

[20] Qasim M, Khan ZH, Lopez RJ, Khan WA. Heat and mass transfer in nanofluid thin film over an unsteady stretching sheet using Buongiorno's model. European Physical Journal–Plus. 2016;**131**:1-16

[21] Wang CY. Liquid film sprayed on a stretching cylinder. Chemical Engineering Communications. 2006;**193**:869-878

[22] Noor SK, Gul T, Islam S, Khan I, Aisha MA, Alshomrani AS. Magnetohydrodynamic nanoliquid thin film sprayed on a stretching cylinder with heat transfer. Applied Sciences. 2017;**271**:1-25

[23] Liao SJ. The proposed homotopy analysis method for the solution of nonlinear problems [PhD thesis]. Applied Mathematics and Computation 147 (2004) 499-513. Shangai Jiao Tong Universit; 1992

[24] Liao SJ. An explicit, totally analytic approximate solution for blasius viscous flow problems. International Journal of Non-Linear Mechanics. 1999;**34**:759-778

[25] Liao SJ. Beyond Perturbation: Introduction to the Homotopy Analysis Method. Boca Raton: Chapman and Hall, CRC; 2003

[26] Liao SJ. On the analytic solution of magnetohydrodynamic flows of non-newtonian fluids over a stretching sheet. Journal of Fluid Mechanics. 2003;**488**:189-212

[27] Abbasbandy S. The application of homotopy analysis method to nonlinear equations arising in heat transfer. Physics Letters A. 2006;**360**:109-113

[28] Alshomrani AS, Gul T. The convective study of the Al_2O_3-H_2O and Cu-H_2O nano-liquid film sprayed over a stretching cylinder with viscous dissipation. European Physical Journal–Plus. 2017;**132**:495

[29] Gul T. Scattering of a thin layer over a nonlinear radially extending surface with magnetohydrodynamic and thermal dissipation. Surface Review and Letters. 1850123. DOI: 10.1142/S0218625X18501238

[30] Wang CY. Fluid flow due to a stretching cylinder. Physics of Fluids. 1988;**31**:466-468

[31] Ishak A, Nazar R, Pop I. Uniform suction/blowing effect on flow and heat transfer due to a stretching cylinder. Applied Mathematical Modelling. 2008;**32**:2059-2066

Analysis of Porous Behaviors by Water Flow Property of Geonets by Theoretical Simulation

Han-Yong Jeon

Additional information is available at the end of the chapter

http://dx.doi.org/10.5772/intechopen.85951

Abstract

The relationship of compressive behavior according to manufacturing process parameters of geonet was investigated. We analyzed the drainage behavior of the bi- and tri-plane geonet used for the planar drainage and investigated the changes of the drainage behavior due to the restraining load. The data showed that there is no critical manufacturing factor to affect the compressive strength of the bi-planar geonet. All of these parameters can affect in a very complicated way. And the strand inclination can mainly affect to the after compressive strength, i.e., roll-over behavior. The results considering site-specific conditions of the landfill system explained the temperature has influence on the compressive behavior of the geonet. The compressive strength was reduced and the strain at yield increase gradually with the temperature for both of bi- and tri-planar geonet. Significant reductions in flow capacity were observed for the traditional bi-planar and cylindrical type geonet and this value was consistent to the compressive strength. These decreases were anticipated due to the abrupt thickness decrease of the geonet caused by roll-over. In the other hand, there was no significant decrease of transmissivity for the tri-planar geonet which has no roll-over phenomena.

Keywords: geonets, compressive behaviors, transmissivity, strand inclination, flow capacity, roll-over phenomena

1. Introduction

The application goal of geonets is mainly for planar drainage, and it is used as a medium to safely discharge the leachate generated from the landfill to the outside of the landfill. The two strands are cross-bonded to form a network structure in geonet, which serves as a drainage passage for the liquid. Geonet products of various structures are used and drainage capacity of geonet is very important because the landfill service life is determined by geonets drainage capacity when designing landfill.

IntechOpen

Fannin [1] investigated the factors influencing the drainage capacity of geonets, and confirmed that the flow rate depends on the structure of the geonet and the drainage capacity is smaller when the strand channel and the crossing angle of the geonet are large.

Also, Zhao [2] analyzed the factors affecting the drainage performance of geonets and confirmed that the drainage capacity decreases as the hydraulic gradient and the compressive strength increase.

To investigate the behavior of geosynthetics under high normal stresses, Narejo and Rad et al. conducted a transmissivity test with normal stresses up to 2000 kPa [3–5]. From the results showed the variation of geonet transmissivity with normal stresses. Geonets behaved similarly up to 200 kPa regardless of weight. However, in the case of geonet with a small strand cross-section area thickness, the drainage performance decreases when the compressive strength increases, but the drainage performance does not decrease much even when the maximum confining load is 2000 kPa for thicker geonets. This is probably due to the fact that as the thickness of the strand cross-section layer constituting the geonet becomes smaller and the compressive strength becomes larger, the roll-over of the strand intersection of the geonet occurs more often.

Koerner [6] analyzed the roll-over phenomenon of bi-planar geonet. According to this, since the upper and lower strands constituting the geonet are not vertically bonded to each other at the cross-section areas, when the compressive strength is applied to the upper and lower layers of the geonet, the drainage capacity is large at the initial stage, the deformation of the strand joint starts to occur. This roll-over phenomenon causes a change in the drainage capacity in the upper and lower layers of the geonet, and the drainage capacity of the geonet is decreased and the drainage performance is lowered with time.

Allen [7] analyzed the effect of orientation of strand constituting geonet on drainage performance. In the case of the lower strand with small inclination angle at the cross-section area of the upper and lower strands, the roll-over phenomenon occurs more than the upper strand, and the drainage performance deteriorates.

Kopp reported that geonet drainage performance was affected by the construction site temperature when the geonet was constructed at the landfill site [8]. Pegg indicates that the drainage performance of the geonet is most affected under the circumstance at 80 or 85°C in the landfill site [9].

Therefore, when the environmental temperature is increased, the mobility of the molecular chain of the polyethylene, which is the raw material of geonet, becomes larger, and the drainage performance may be decreased due to the structural change of the geonet to the compressive strength.

In this chapter, the variation of drainage performance of geonets used for horizontal drainage is analyzed in relation to the influence of constraint load related factors.

2. Materials and performance test

2.1. Specimens

Three samples of geonets were used. The first geonet is a 5.6 mm thick high density polyethylene (HDPE) traditional bi-planar geonet. A photograph of this sample is shown in **Figure 1**.

(a)   (b)   (c)

Figure 1. Photograph of various geonets samples: (a) bi-planar, (b) tri-planar, and (c) cylindrical type.

The second and third geonet are also HDPE, and have 8.6 mm tri-planar, 8.2 mm cylindrical type bi-planar structure, respectively. All these material are used for landfill cover and lining system drainage. Typical specifications for samples are provided in **Table 1**.

2.2. Short-term compressive test

Short-term compressive deformation testing was performed in accordance with ASTM Standard Test Method (ASTM D6364-06; Standard Test Method for Determining Short-Term Compression Behavior of Geosynthetics) [10].

The geonet specimens were placed between two steel sheets and compressed at a strain rate of 1.0 mm/min. The compressive strength was measured while varying the strain. The compressive strength of the geonet is determined by parameters such as thickness, mass per unit area, crystallinity, strand strength, bonding strength of strand, strand angle and inclination in plane, test conditions are 23, 35 and 50°C, Strain rates were 0.1, 0.5, 1.0, 5.0 and 10 mm/min. **Figure 2** shows the test pattern according to various parameters.

2.3. Transmissivity test

The horizontal permeability, transmissivity, which is the drainage performance of the geonet, was measured using the ASTM Standard Test Method (ASTM D4716/D4716M-14; Standard Test Method for Determining the In-plane Flow Rate per Unit Width and Hydraulic Transmissivity of a Geosynthetic Using a Constant Head) [11]. And this is determined from the relationship between the number of paths per unit area and the in-plane drainage capacity, the compressive strength and the hydraulic gradient.

The short-term transmissivity test of geonet was carried out under various normal stresses and three hydraulic gradient conditions at 0.1–1.0, and the test temperature was 22–23°C. **Figure 3** shows a schematic diagram of the transmissivity test apparatus and the transmissivity value was calculated from the following equation:

$$\theta = \frac{Q}{B \times (\Delta h/L)} \qquad (1)$$

where θ = transmissivity (m²/s), Q = volume of discharged fluid per unit time (m³/s), L = length of the specimen (m), B = width of the specimen (m) and h = difference in the total head across the specimen (m).

Property	Test method	Unit	Bi-planar	Tri-planar	Cylindrical type
Thickness	ASTM D5199	mm	5.6	8.6	8.2
Mass per unit area	ASTM D5261	g/m^2	920	1700	2300
Carbon black	ASTM D4218	%	2.3	2.2	2.3
Density	ASTM D1505	g/cm^3	0.942	0.944	0.940
Crystallinity	ASTM D2910	%	56	55	61

Table 1. Basic properties of various geonet samples.

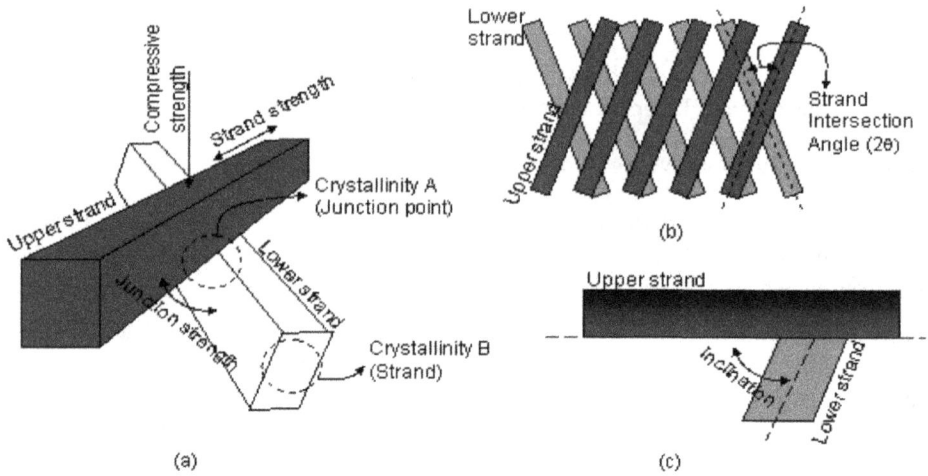

Figure 2. Schematic diagram various test parameter for determining the main factor: (a) 3-D, (b) horizontal, and (c) verticle.

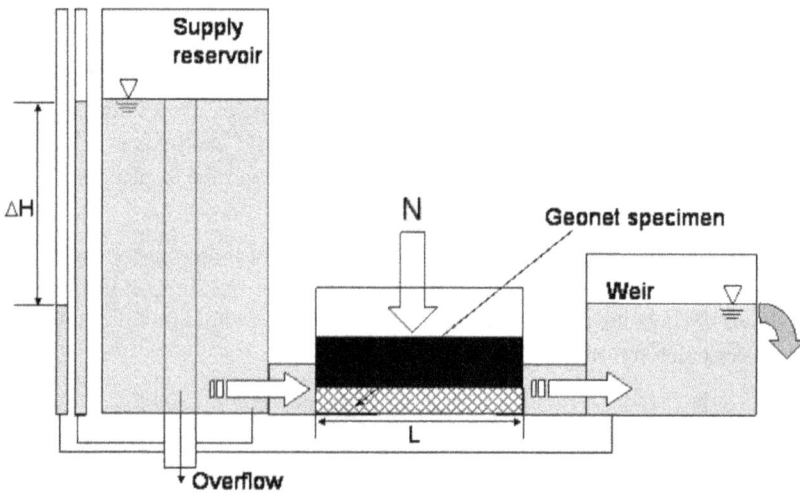

Figure 3. Schematic diagram of transmissivity test device.

3. Results and discussion

3.1. Factors affecting the short-term compressive behavior

Figures 4–10 show the results of correlation between various manufacturing parameters and the compressive strength. The relationship between all of these manufacturing parameters of geonet and the compressive strength were not well defined through analysis of experimental values and it is assessed that the compressive strength of a geonet cannot be related to such as weight, geometrical properties, crystallinity, rib mechanical properties, etc. Only, compressive strength is determined by a combination of these factors and thus has to be controlled as any other material constant of a geonet.

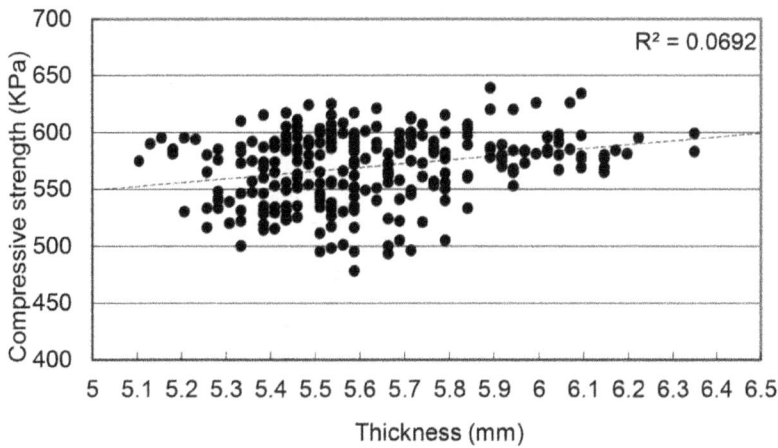

Figure 4. Plot of compressive strength vs. thickness.

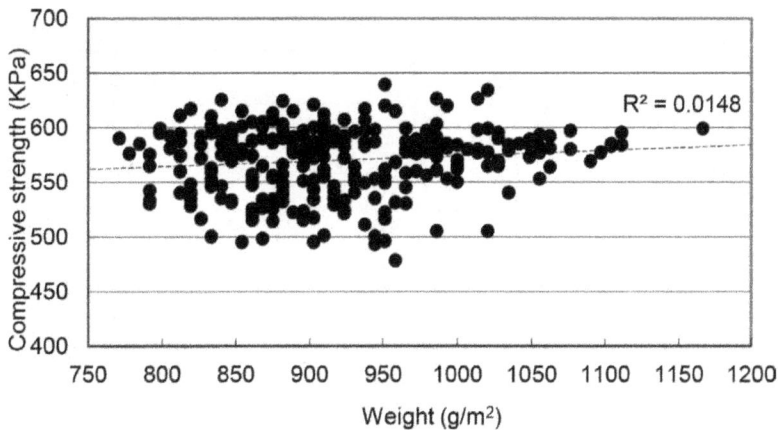

Figure 5. Plot of compressive strength vs. mass per unit area.

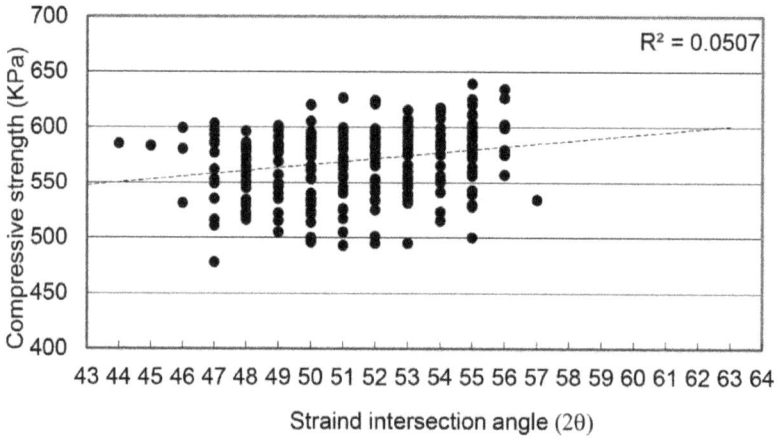

Figure 6. Plot of compressive strength vs. strand intersection angle.

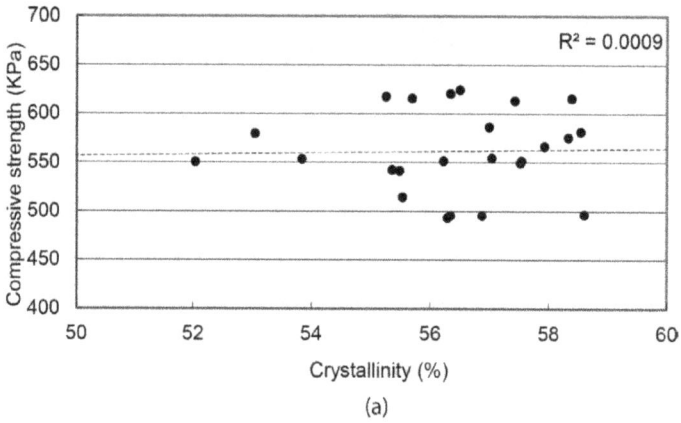

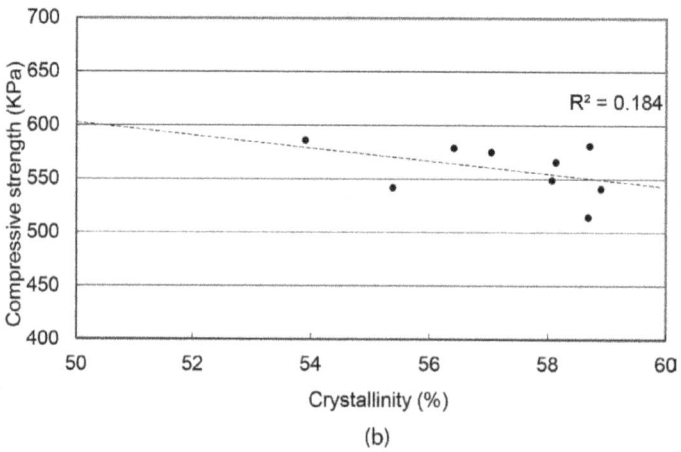

Figure 7. Plot of compressive strength vs. crystallinity: (a) at intersection and (b) at strand.

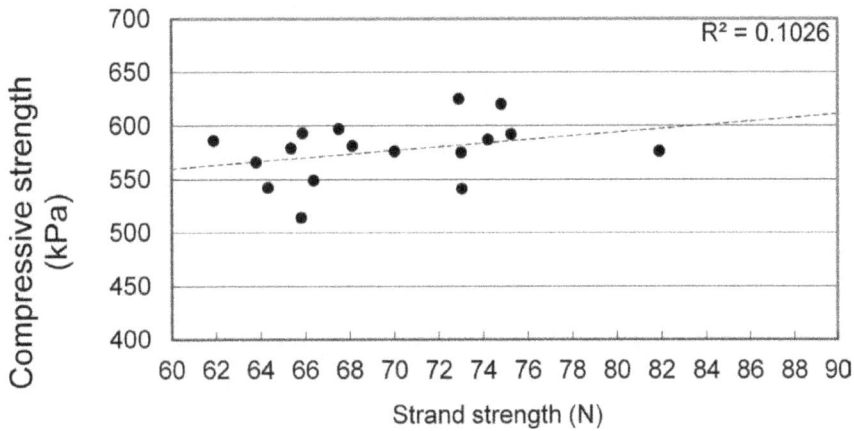

Figure 8. Plot of compressive strength vs. strand strength.

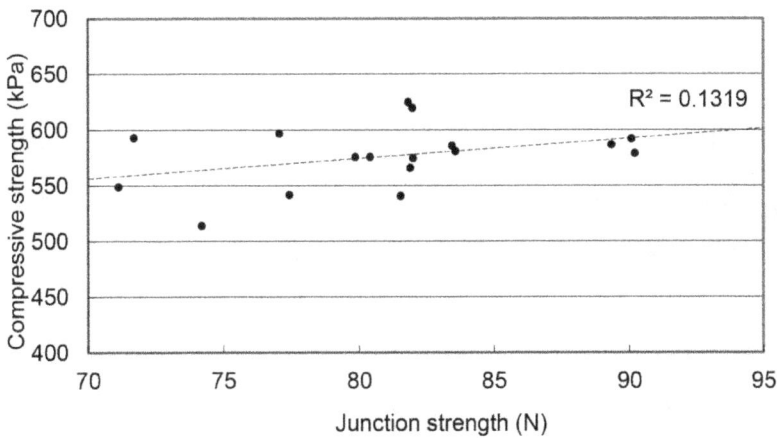

Figure 9. Plot of compressive strength vs. junction strength.

To determine the compressive strength, the strand slope associated with the range of roll-over may not affect the confining strength, since the confining strength measurement time is after the start or end point of the roll-over (**Figure 11**). In **Figure 11**, the roll-over distance of a rect-angular geonet strand is longer than that of a circular stranded geonet. Additional evidence for the relationship between the strand cross-sectional shape and the rollover behavior to support this can be seen by comparing the shape of the compressive strength-strain curve of the geonet shown in **Figure 12**.

The cylindrical type exhibits a very stiff behavior even after the end of the roll-over and has very short roll-over range compare to the bi-planar which has very long roll-over region. By defining the factor that affects to the short-term compressive strength behavior of the bi-planar geonet, it should be determined that before the determining point of compressive

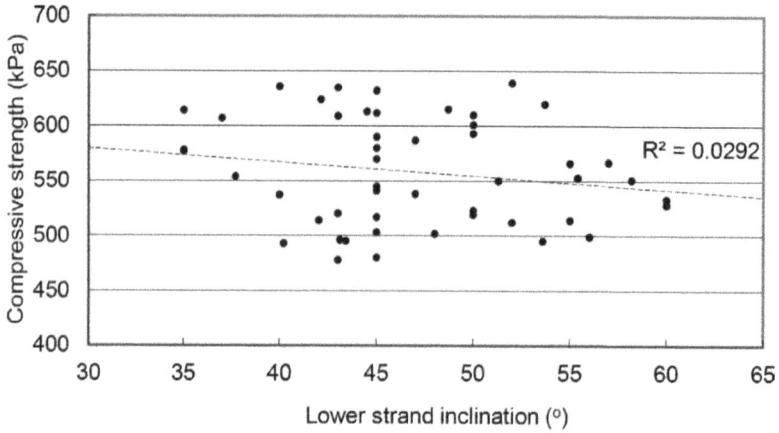

Figure 10. Plot of compressive strength vs. strand inclination.

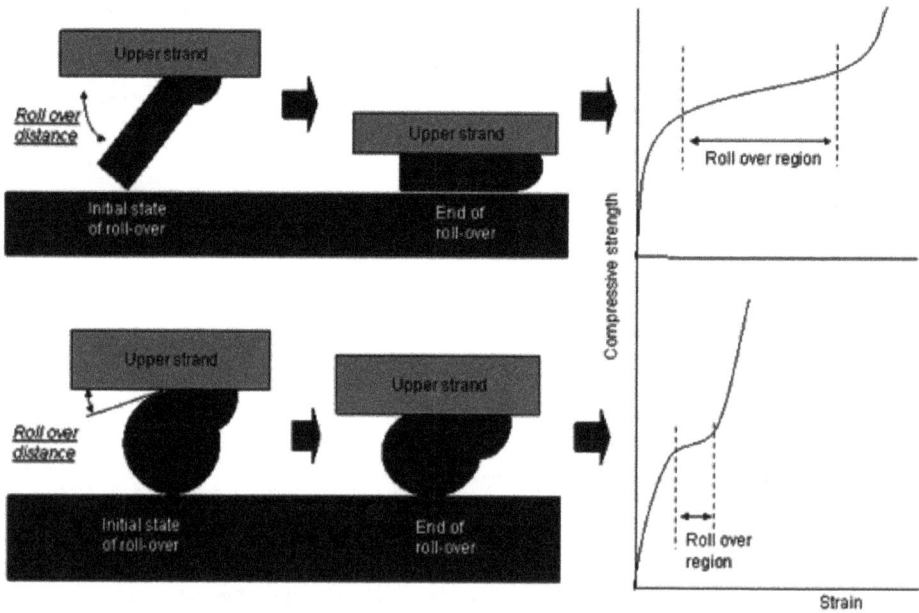

Figure 11. Relationship between strand cross sectional shape and roll-over range.

strength. Also, all of the manufacturing parameters can affect the transmissivity of the geonet and strand inclination and strand cross-sectional shape will affect the roll-over phenomenon mainly.

Therefore, it is concluded that if the cross-sectional shape of the strand is close to the circle, the roll-over region could be shortened dramatically and this advantage will contribute to the advance the long-term flow capacity of the geonet.

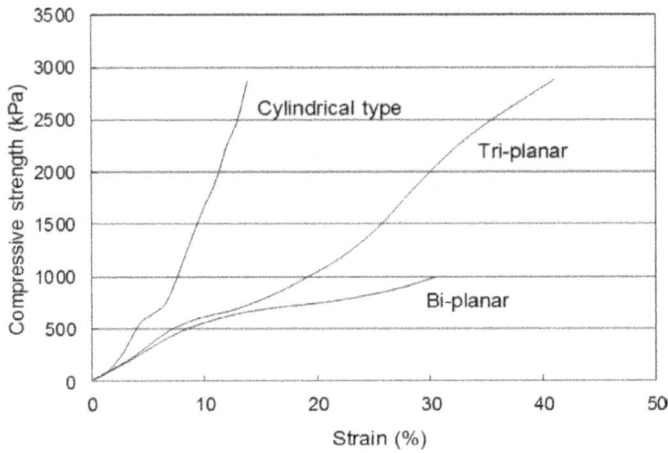

Figure 12. Plot of compressive strength-strain curves of various geonet samples.

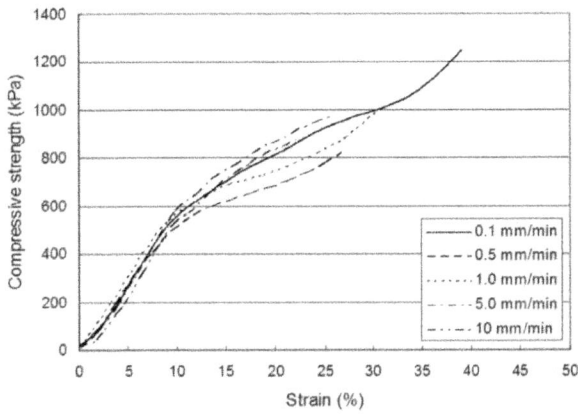

(a)

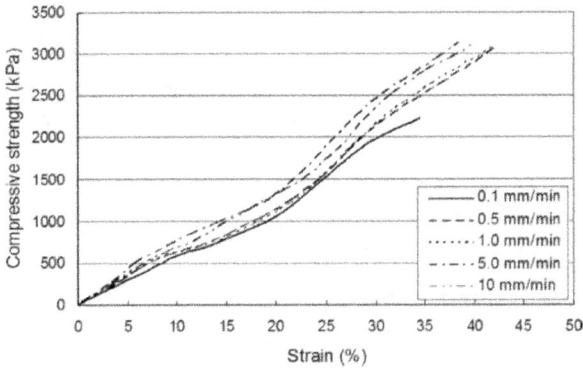

(b)

Figure 13. Plot of compressive behavior curves according to various deformation rates at 23°C test temperature: (a) bi-planar and (b) tri-planar.

3.2. Short-term compressive behavior

Figures 13–15 show all the results of compressive strength-strain curves at different temperatures (23, 35 and 50°C) and elongation rate (0.1, 0.5, 1.0, 5.0 and 10 mm/min).

In here, the compressive strength decreases and the deformation value increases according to the temperature. And the initial slope of this curve decreases greatly with increasing temperature, and the elastic modulus with temperature increases in the given strain range.

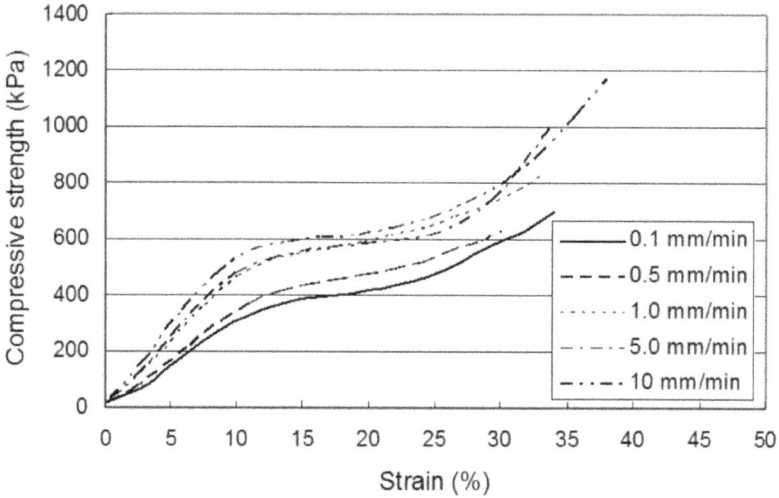

(a)

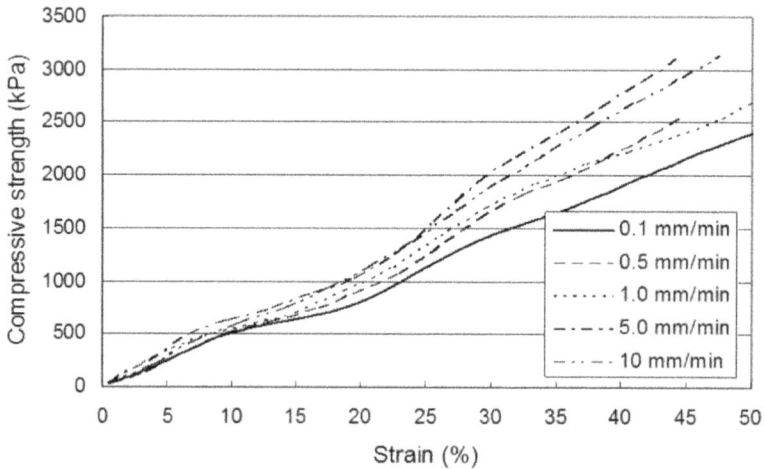

(b)

Figure 14. Plot of compressive behavior curves according to various deformation rates at 35°C test temperature: (a) bi-planar and (b) tri-planar.

(a)

(b)

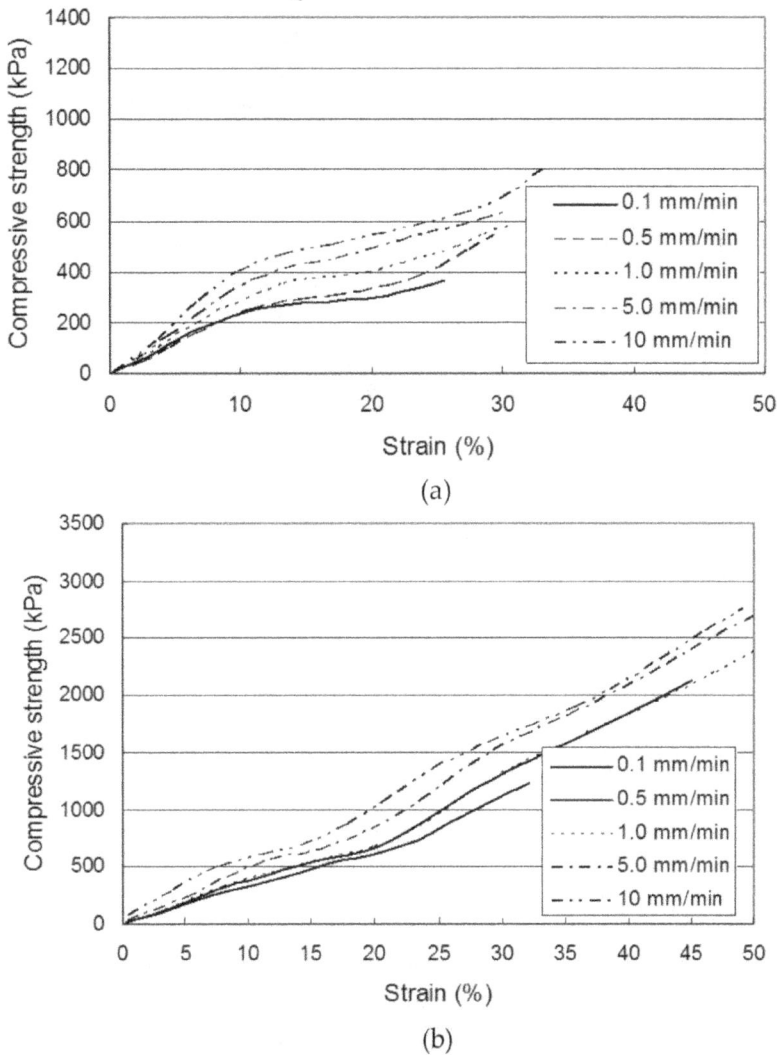

Figure 15. Plot of compressive behavior curves according to various deformation rates at 50°C test temperature: (a) bi-planar and (b) tri-planar.

Considering the landfill's severe temperature conditions, this decrease may be affect the long-term flow capacity of the geonet drainage.

3.3. Short-term transmissivity

The flow capacity data as logarithm of transmissivity as a function of normal pressure (log value) is presented in **Figure 16**. This type of plot more clearly reveals the material response to applied loading conditions. **Figure 16** shows a continuous decrease in transmissivity from

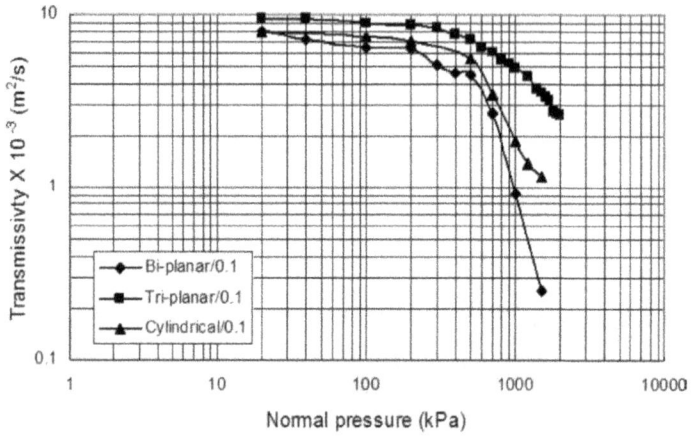

(a)

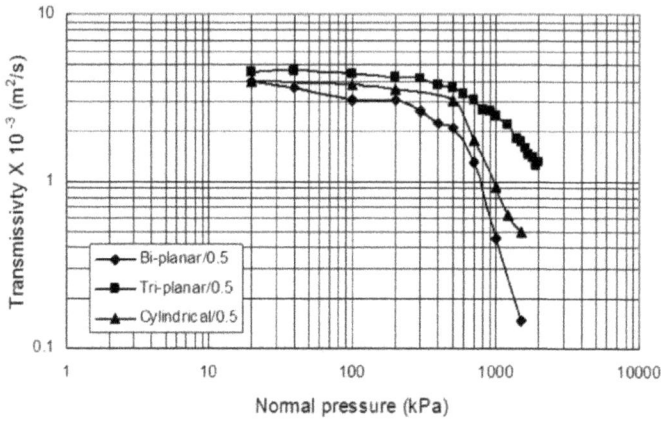

(b)

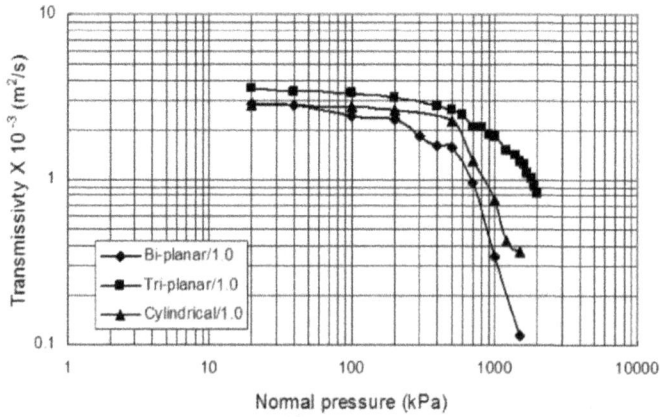

(c)

Figure 16. Plot of transmissivity results of various geonet samples: (a) gradient−0.1, (b) gradient−0.5 and (c) gradient−0.5.

8.2 to 0.25, 9.53 to 2.62 and 8.2 to 1.16 (×10^{-3}) at 0.1 hydraulic gradient for bi-planar, tri-planar and cylindrical type geonet, respectively. Also in other hydraulic conditions show same trend as 0.1 hydraulic gradient. It is clear from **Figure 16** that bi-planar and cylindrical type show dramatic decrease of transmissivity above 600 kPa.

However, the cylindrical type geonet has more strong flow capacity compare to the bi-planar geonet even though there is a roll-over effect. On the other hand, the tri-planar geonet shows the excellent resistance to the thickness decrease.

The transmissivity of the tri-planar geonet is very high but decrease gradually with increasing normal pressure (compressive strength). This is due to the tri-planar geonet structure and means the stability of tri-planar geonet even though very high normal pressure (2000 kPa) and hydraulic gradient (1.0).

Therefore, it can be concluded that the triplanar geonet drainage pattern is linear and that a tri-planar geonet with a zig-zag-shaped flow pattern can discharge the liquid very effectively even when the thickness is the same as the bi-planar geonet.

From these results, it is considered that the most important factors affecting short-term flow capacity of geonet are drainage path and drainage pattern (e.g., linear or zigzag type) and geonet thickness.

4. Conclusion

The data from short-term engineering properties describe the influencing factors to the short-term compressive for the bi-planar geonet, compressive behavior under various test conditions for the bi-planar and tri-planar geonet, and the short-term flow capacity for three types of geonet under up to 2000 kPa normal pressure.

The following conclusions are drawn:

1. The compressive behavior of bi-planar geonet shows different drainage behavior before and after compression. Compared with the pre-compression, it can be affected by the strand cross-pattern after compression, and the strand inclination and cross-sectional pattern affect the roll-over behavior, resulting in a larger reduction in drainage performance.

2. Since these materials are viscoelastic in nature, compressive behavior was affected by the temperature changes. The compressive decreases were up to 40% when compared to its value for 23°C for both bi- and tri-planar geonet in very critical temperature condition (50°C).

3. The reduction of the drainage capacity depends on the structure of the geonet, and the drainage pattern is composed of a zigzag pattern rather than a straight line, or a smaller triple-planar geonet with a larger strand thickness. In this case, the roll-over effect can be minimized, and even when the compressive strength is higher than 600 kPa, the decrease of drainage capacity is smaller than that of the bi-planar geonet. And the selection of a geonet to maximize flow capacity in the long-term must consider not only the thickness of the strands but also the pattern of the liquid flow channel.

Author details

Han-Yong Jeon

Address all correspondence to: hyjeon@inha.ac.kr

Department of Chemical Engineering, Inha University, Incheon, South Korea

References

[1] Fannin RJ. Factors influencing the long-term flow capacity of geonets. In: 3rd International Conference on Geosynthetics. 1995. pp. 267-280

[2] Zhao A. Effect of soil presence on flow capacity of drainage geocomposites under high normal loads. In: 7th International Conference on Geosynthetics. 1999. pp. 799-812

[3] Narejo D. Long-term performance considerations for geonet drainage geocomposites. In: 57th Canadian Geotechnical Conference, Session 6D. 2005. pp. 12-15

[4] Narejo D. Design strength for geonet geocomposite. In: Proceedings of the Geoysnthetics 2007 Conference. 2007. pp. 71-78

[5] Rad NS, Gokmen C, Stalcup JM. Factors affecting hydraulic transmissivity of geocomposite drain systems. In: 6th International Conference on Geosynthetics 1998. pp. 1055-1058

[6] Koerner RM. Designing with Geosynthetics. New Jersey, USA: Prentice Hall Inc; 2005

[7] Allen S. Measuring the strand inclination angle of a bi-planar geonet. In: Proceeding of the Geosynthetics 2007 Conference. 2007. pp. 79-92

[8] Kopp B. Membrane sealing for an irrigation canal KIRKUK IRAQ. In: Geosynthetics Case Histories: International Society for Soil Mechanics and Foundation Engineering. 1993. pp. 48-49

[9] Peggs ID, Lawrence C, Thomas R. The oxidation and mechanical performance of HDPE geomembrane: A more practical durability parameter. In: Proceedings of the 7th ICG. 2002. pp. 779-782

[10] ASTM D6364-06. Standard test method for determining short-term compression behavior of Geosynthetics. In: Annual book of ASTM standard. Pennsylvania, USA: ASTM; 2018

[11] ASTM D4716/D4716M-14. Standard test method for determining the (in-plane) flow rate per unit width and hydraulic transmissivity of a geosynthetic using a constant head. In: Annual Book of ASTM Standard. Pennsylvania, USA: ASTM; 2014